A Delicious History of Kyoto:
The Geographical History of Restaurant Culture in Japan's Former Capital

A Delicious History of Kyoto:
The Geographical History of Restaurant Culture in Japan's Former Capital

—

Masahiro Kato
Naomi Kawasumi

UNIVERSITY PRESS OF
RITSUMEIKAN

UNIVERSITY PRESS OF
RITSUMEIKAN

8 Nishinokyo-Higashi-Togano-cho
Nakagyo-ku
Kyoto 604-8520
JAPAN

The University Press of Ritsumeikan is part of Ritsumeikan University. It furthers the University's mission by disseminating knowledge in the pursuit of education, research, and scholarship at the global levels of excellence.

This publication is translation of *Oishii Kyoto Gaku* published by Minerva Shobō, Kyoto, Japan, 2022.

First English edition, 2025

ISBN: 978-4-9913516-0-0 (PDF)
978-4-9913516-1-7 (Print Edition)

Published by the University Press of Ritsumeikan with the Cooperation of Sciendo (De Gruyter Poland Ltd, Warsaw) / Part of Walter de Gruyter GmbH, Berlin/Boston
10.2478/9784991351600-fm

10.2478/9784991351600-toc

Contents

10.2478/9784991351600-001

List of Illustrations

10.2478/9784991351600-002

Author Biographies

Masahiro Kato
Born in Nagano prefecture, Masahiro is a faculty member of the Faculty of Letters at Ritsumeikan University. He graduated from the Faculty of Humanities at the University of Toyama, completed the doctoral program at the Graduate School of Letters at Osaka City University, and received a doctorate in literature. His specialty is human geography. His major publications include *Sakaba no Kyoto gaku* (Studies of Kyoto Bars) (Minerva Shobō, 2020) and *Osaka: Toshi no kioku o hori okosu* (Osaka: Unearthing Memories of the City) (Chikuma Shinsho, 2019).

Naomi Kawasumi
Born in Osaka Prefecture, she is a faculty member of the Faculty of Letters at Ritsumeikan University. She graduated from the Faculty of Letters at Ritsumeikan University, completed the doctoral program at the Graduate School of Letters at Ritsumeikan University, and received a doctorate in literature. She specializes in geography. Her major papers include “Kindai Kyoto ni okeru kōgai jyūtakuchi no kaihatsu to sono ricchi tokusei: Omuro Komatsuno-chō o rei to shite” (Development and Situational Characteristics of Suburban Residential Areas in Modern Kyoto: A Case study of Omuro Komatsuno-chō) (Ritsumeikan Bungaku, 2020) and “Kindai Kyoto ni okeru tochi riyō to chikei kankyō” (Land Use and the Geomorphic Environment of Modern Kyoto) (*Ritsumeikan Pan-Pacific Civilization Studies*, 2019).

10.2478/9784991351600-003

Acknowledgments

This book was supported in part by a JSPS Grant-in-Aid (16H01965 Representative: Keiji Yano), and the authors would like to express their gratitude for this invaluable support. We would also like to thank Itaru Wakui (editor of the Minerva Shobō), who took an early interest in the author's (Kato's) website following *Studies of Kyoto Bars* and took charge of editing this book.

10.2478/9784991351600-004

Foreword

What do you think of when you hear the phrase "Kyoto cuisine"? This book is an attempt to retrace the cultural genealogy of the cuisine and restaurants of modern Kyoto, by walking through its urban history from the Edo period up to the present day.

The genre of Kyoto cuisine is firmly established, and we will not examine it here. This book instead covers the role of geography in the development of various dishes which stand out as specialties of Kyoto cuisine, including: *imobō*, saltwater and freshwater fish dishes, eel dishes, chicken dishes, and Western-style dishes. Except for *imobō*, these dishes are by no means unique to Kyoto. However, these commonplace dishes, which are seemingly everywhere, conceal an unlikely history. For example, why is "Tokyo-style" (Edo-yaki) broiled eel served in the eel restaurants of Kyoto? This book takes the reader on a journey to explore the roots (origins and routes) of the cuisines that have become so deeply integrated into the foodscape of present-day Kyoto.

10.2478/9784991351600-005

Foreword to the English Version

In December 2013, "*washoku*: traditional dietary cultures of the Japanese" was inscribed on UNESCO's List of Intangible Cultural Heritage. Although the international appreciation of Japanese cuisine considerably predates this, there is no doubt that *washoku* has come to be greatly enjoyed by visitors to Japan. At the same time, there is growing academic interest in food, with Ritsumeikan University establishing the College of Gastronomy Management in April 2018, followed by the Department of Japanese Food Culture at Kyoto Prefectural University's Faculty of Letters in April 2019. A broad range of texts have been published approaching the subject of food from various angles.

Located in the Japanese archipelago off the eastern edge of the Eurasian continent, Kyoto – a city whose history spans more than a thousand years – is home to *kyō-ryōri* (Kyoto cuisine), which is counted among the most iconic of *washoku*. Central Kyoto is packed with restaurants featuring the words "*kyō-ryōri*" on their signboards, but this single phrase fails to capture the truly diverse cuisine and food culture that has developed in Kyoto. Beginning in the Meiji era (1868–1912), when Japan opened to the rest of the world after centuries of isolation throughout the preceding Edo period (1603–1868), a unique culinary culture was created by selectively incorporating foreign foods and techniques and combining ingredients and cooking methods from the diverse regions of Japan. The mother of this culinary culture, perhaps unsurprisingly, was the restaurant.

This book is a historical-geographical attempt to retrace the genealogy of restaurants and their culinary culture in Kyoto, focusing on several culinary cultures born and rooted in Kyoto, travelling back and forth from the Edo period to the present. Drawing on tourist guidebooks, literary works, newspaper advertisements, maps, illustrations, old photographs, biographical records, and even telephone directories, we explore the foodscape that has blossomed in modern Kyoto and unearth the origins of culinary cultures that have been hidden in the fringes of its urban spaces.

We start by shining a light on an amusement park which has been long forgotten in modern Kyoto. Yase Amusement Park, located on the northeastern outskirts of the city, was once home to many of Kyoto's best-known restaurants, creating a space resembling a culinary theme park. This theme park was symbolized by a restaurant called Hiranoya, located on the banks of a pond at the park's center, renowned for a unique dish called *imobō*. Chapter 1 explores the *imobō* and Hiranoya's expansion into multiple locations.

Imobō is prepared by skillfully cooking taro, which originates from Nagasaki, and *bōdara* dried cod, a seafood product from Hokkaido. Despite being an inland city, many *kyō-ryōri* dishes feature seafood. Typical examples include *nishin soba* (herring with buckwheat noodles) and *sabazushi* (mackerel sushi). These dishes are complemented by the diverse kinds of freshwater fish caught in the inland waters in and around the city, including the swift waters of the Kamo River that flow through the city center. Chapter 2 discusses these freshwater fish, which often make it to the table in dishes that are unique to this inland city.

Across Japan, for many people, eel is the first thing that comes to mind when they think of freshwater fish (setting aside the migratory habits of eel, which take it to the oceans). To this day, on midsummer's day, when eel is traditionally eaten to counteract the humidity of the Japanese summer, you can still see fragrant smoke rising from eels being grilled over charcoal at freshwater fish restaurants. The reader may be aware that there are two main cooking (grilling) methods for eel. This is often referred to as a cultural difference between Tokyo and Osaka. However, although Kyoto is in Kansai, the same region as Osaka, for some reason, the Tokyo style of grilling is more commonly used in Kyoto's restaurants. To understand how this came about, Chapter 3 examines the Fourth National Industrial Exhibition, a major event held in Kyoto in the mid-Meiji period and explains how eel cuisine developed in Kyoto.

In prewar modern Japan, major events such as the Fourth National Industrial Exhibition created opportunities for urban culture to blossom. The ascension of the Japanese emperor to the throne in 1915 was another significant event in the development of Kyoto cuisine. At the time, with all of Japan in a celebratory mood, Shin Miura, a restaurant specializing in chicken hotpot with humble origins in the old red-light district of Hakata in Kyushu set up shop without much fanfare on the outskirts of the old city area. Who could have predicted back then that the Shin Miura style of hot pot cuisine would later take over Kyoto, sweeping aside the city's traditional cooking method (*sukiyaki*)? Chapter 4 traces the lineage of *mizudaki*, a chicken dish that has now become so well-established that its origins outside Kyoto have been largely forgotten.

Moving forward to the Showa era (1926–1989), although it is hard to imagine now, when the imperial ascension was held in Kyoto in 1929, chicken was a more prized luxury item than beef. And it was around this time that Western-style restaurants serving pork, beef, and chicken began to open in downtown Kyoto. These Western-style restaurants naturally served steak, but also Japanese interpretations of Western-style foods such as croquettes, pork cutlets, and curry with rice. Although Kyoto is often associated with *washoku*, Chapter 5 reveals how the thriving culinary scene inspired Kyoto to develop into one of Japan's foremost Western food-loving cities.

In Chapter 6, the authors share some interesting stories they discovered in writing this book which serve to demonstrate the centrality of Kyoto's restaurants and food culture to the city. This book provides an immersion in the multifaceted history of the foodscape of Kyoto. But seeing is believing, and this holds true for tasting, so nothing beats a visit to Kyoto to enjoy these dishes.

Written by Masahiro Kato and Naomi Kawasumi in autumnal dusk, 2023

10.2478/9784991351600-006

Explanatory Notes

The main databases used to search and browse sources and newspaper articles were the following: National Diet Library Digital Collections; Kikuzo II Visual (Asahi Shimbun Database) [now Asahi Shimbun Cross Search]; Maisaku (Mainichi Shimbun Database); Yomiuri Database Service (Yomiuri Shimbun Database).

The following measures have been taken to aid readability when quoting from historical sources. Additions and observations by the authors (citers) are shown in square brackets [ ] and emphasized with underlining where necessary. The quoted texts may contain some terms and expressions that are considered inappropriate today, but the original text has been retained because of its value as a historical source.

The book refers throughout to the "Modern Kyoto Overlay Map" published online by the Art Research Center of Ritsumeikan University (https://www.arc.ritsumei.ac.jp/archive01/theater/html/ModernKyoto/). This resource allows you to select a historical map and view it as an overlay to modern-day Google Maps, moving the slider at the top right of the map screen to the left or right to change the transparency of the overlay. The reader can select different maps, both past and present, and refer to them as you read this book.

10.2478/9784991351600-007

Introduction
Welcome to the Culinary Theme Park

Walking a short way along a path lined with the overhanging branches of old maple trees, we came upon a rustic gateway, with a weathered sign that read "Heihachi Jaya." An elegant bell with a wooden clapper hung beside it. Iki paused.

"This looks like quite an old place, doesn't it?"

"Yes, this area used to be a settlement on the Wakasa Highway that ran to Kyoto from the coast of the Sea of Japan. This place is a roadside teahouse and inn that has been around since the Edo period. I hear that it's long been renowned for its freshwater fish dishes."[1]

1. To the noble summit of Mount Hiei!

A guidebook published in the early Shōwa period advises: "First and foremost, let us go to the noble summit of Mount Hiei and to the enchanted land of Yase!"[2]

One may wonder whether the ancient samurai lords in the tower of Osaka Castle ever caught sight of Mount Hiei just beyond Kyoto (about 50 kilometers away and 850 meters above sea level) when surveying their territory. Today, it would take about 70 minutes to get from Kyōbashi Station (on the Keihan Electric Railway), located north of Osaka Castle Park, to Yase-Hieizanguchi Station (on the Eizan Electric Railway) at the foot of the mountain, albeit with a change of trains at Demachiyanagi Station.

The writer Kunio Yanagita once called railroads (or rather, steam trains) "a technique for shrinking the Earth."[3] With the incremental shortening of time and distance that accompanied the development of transportation systems, the sacred peak of Mount Hiei became a sightseeing destination that could be enjoyed on a day trip from Kyoto, Osaka, or Kobe in the prewar Shōwa era (1926–1939). The Keihan Electric Railway opened the Ōtō Line, which extended to Demachiyanagi in 1989, but the Eizan Electric Railway is much older.

Zensuke Ōsawa (1854–1934), the founder of the Eizan Electric Railway, explained the background to these developments:

> The prosperity of Kyoto has shifted from the east to the south, and although there have recently been signs of an impetus toward development to the west, the north remains much as it has always been. This is mainly due to the lack of public transportation. There have been at least two or three plans to lay electric railways in the north. The

> Kyoto Electric Railway Company had already obtained government permission to reclaim the Kamo River to the north of Imadegawa and launch an electric railway to Yase but abandoned the project halfway through. Sakichi Yoshida also launched the construction of a steam railway from Demachiyanagi to
>
> Shimeigatake on Mount Hiei, and this too had been approved by the government but was also abandoned midway through. There were quite a few other similar hopefuls, but only few had much prospect of seeing it through to the end.[4]

Zensuke Ōsawa was a founder of the Kyoto Electric Light Company and a central figure in the Kyoto business world. He noticed a geographical imbalance in the urban development of modern Kyoto and made large investments of capital in undeveloped areas. Although similar plans had foundered, the Kyoto Electric Light Company with Ōsawa at the helm made a major contribution to opening the northern area of Kyoto for development with the construction of the Eizan Electric Railway.[5]

Beginning with a railway line running across the flat country between Demachiyanagi and Yase (opened in September 1925), they then built a cable car to the highest point of Yase at the foot of the mountain (Shimeidake, opened in December 1925) and finally a ropeway (referred to as the "overhead cableway") that crossed the valley between Shimeidake and Enryakuji (opened in September 1928). By combining these three types of modern transportation, the summit of Mt Hiei became directly accessible from urban Kyoto (Fig. 0.1). Commenting on the ski resort that would later open at Shimeidake, the Kyoto Electric Light Company boasted that "you can enjoy night skiing only 30 minutes after dining within Kyoto city."[6]

Figure 0.1 Cable car

In this way, a transportation network was constructed that made tours of the Eizan area possible, including the Ōmi (Lake Biwa) side of the area, and a new sightseeing space, referred to as "Rakuhoku" (an alternate historical name for northern Kyoto), was born in the suburbs of the historic city of Kyoto.[7] An *emaki* picture scroll by Hatsusaburō Yoshida entitled "Bird's Eye View of Sightseeing Transportation for Famous Locations of Mount Hiei" provides a vivid depiction of Rakuhoku (Fig. 0.2).

This now brings us to our next destination, Yase, which is located at the foot of the mountain.

Figure 0.2 View of sightseeing spots drawn by Hatsusaburō Yoshida

2. To the enchanted land of Yase!

At Demachiyanagi, take the main line on the Eizan Electric Railway, which slowly wends its way northward along the left bank of the Takano River.

The route has been ideal for mountain excursions since the railway first opened, with sightseeing spots scattered along its length, including Shisen-dō Temple, Manshu-in Temple, Shugakuin Imperial Villa, and Miyake Hachimangū Shrine.

After passing the environs of Matsugasaki, where the Iwakura River merges with the Takano River. The track follows the latter, curving eastward along its banks. Located at the bottom of a basin, the historic capital of Kyoto received goods from all directions via numerous highways, including the Wakasa Highway (Tsuruga Highway), which originates in Obama in Wakasa, then passes through Kutsuki and Hanaori Pass in Ōmi and Ōhara, where Sanzen-in Temple is located, before reaching Kyoto. The train runs parallel to this road, which is popularly known as the "Saba Kaidō" (Mackerel Road), shortly before arriving at Yase-Hieizanguchi Station.

It is a quiet village. Other than a souvenir shop selling *shibazuke* pickles just outside the ticket gate, there's not much to draw the eye. A hotel with a calm yet luxurious exterior seems to occupy a vast area of land (where the Sports Valley Kyoto used to be), but there's not much of interest here; so, we cross the clear waters of the Takano River and stand on the opposite bank. Despite its venerable age, a wooden building still stands beside a muddy pond. A hall that looks like a guest room can also be seen, so the old wooden building was perhaps once a dining room. Between the pond and the slope of the mountain, we can also see a cleared patch of land where the buildings have been demolished.

There is another building, spread over two floors, facing the Takano River to the side of the stone steps that lead up to the Yase Cable Car Station. With its impressive facade, it must have once also been some kind of catering establishment. Sealed off with ropes, it seems to now be in ruins. This is not a place where one can expect visitors all year round, but only in the high seasons when sightseers come for the spring greenery and autumn leaves.

To the southeast of Yase Cable Car Station, a path weaves up the slopes. The slopes are occasionally interrupted by flat areas of land marked by stone walls, suggesting that buildings once stood there. Traces of waterways can be seen running across these steep slopes, lending them the air of a garden. Further on, a stone commemorating the relocation of the capital to Heian-kyō has for some reason been erected. Are these the remnants of the "enchanted land" of Yase?

Let us now detour slightly, returning to Damachiyanagi Station in the 1920s during the early Shōwa period:

> Alight from the streetcar [at Kawaramachi Imadegawa], head north, cross the Demachi Bridge and one finds Demachiyanagi Station. From here, a comfortably spacious train runs quietly along a flat land to the right side of the Takano River on a straight track. Starting amid the old cedars that grow thickly at Shimogamo Shrine, one passes the imperial university, Yoshidayama, and Daimonjiyama. After going through another six stations – Mototanaka, Chayama, Ichijōji, Shūgakuin, Yamabata, and Miyake Hachiman – one arrives at Yase Station at the foot of Mount Hiei. This is the final station on the flat part of the line, and one feels not a trace of fatigue over the course of the 18-minute journey to get there as the train, equipped with soft seats, moves smoothly along the beautiful scenery outside.
>
> *Passing through the ticket gate at Yase Station, one enters Yase Amusement Park,* which skillfully incorporates a mountain gorge, a river, and two waterfalls into its natural gardens. It was constructed using substantial investments and is fully equipped with all kinds of facilities; it is always packed with visitors as the park is truly worthy of praise.[8]

Today, when you exit Yase-Hieizanguchi Station, you find yourself in a quiet village, but this was once the Yase Amusement Park, which was developed near the station. The company history of the Kyoto Electric Light Company states:

> The amusement park is located in Yase, the final station on the flat line on the banks of the clear waters of the Takano River. This tranquil park covers an area of about 30,000 *tsubo* [approx. a square kilometer] and makes use of the blessed natural beauty with its waterfalls, ponds, and mountain streams. Moreover, the park is dotted with eateries constructed in a rustic style and equipped with a swimming pool, roller skating, electric

> trains, electric vehicles, a zoo, an insectarium, an aquarium, a botanical garden, and all kinds of exercise equipment, quite suitable for washing away the city dust with a day of wholesome enjoyment...[9]

It seems that the amusement park opened in December 1925 when the cable car started running.[10] Despite the chilliness of the season, we know that the amusement park had "a swimming pool, roller skating, electric trains, electric vehicles, a zoo, an insectarium, an aquarium, a botanical garden." A typical amusement park echoes with the joyful shrieks of children playing with playground equipment.

This "tranquil" amusement park, with its "blessed natural beauty," clear rivers, waterfalls, mountain streams, and ponds as well as its "eateries constructed in a rustic style" would play a significant role in the development of Kyoto style cuisine.

3. The *Rakumi* Paradise

The advertisement entitled "The Refreshing Bounty of Yase in Summer" (Fig. 0.3), drawn from a bird's-eye view from the northwest, is useful in gaining a spatial understanding of the Yase Amusement Park, with its peculiar "tranquil" nature.[11] The Takano River gently snakes its way through the foreground, and a large pond is depicted in the center of the map – visitors to the site can still see the small island-like structure in the pond to this day. The Wakasa Highway runs through an area in the front (lower part), just outside that depicted on the map, from the blank area with the characters "Saitōbashi" in the lower left. With the Cable Car Station on the left and Yase Sation on the right, we get a rough idea of the extent of Yase Amusement Park.

Figure 0.3 Bird's-eye view "The Refreshing Bounty of Yase in Summer"

The first thing to note in the layout of the map is that the amusement facilities such as tennis courts, skating rinks, fishing grounds, archery grounds, and swimming pools have all been placed in a single space between the railroad and the Takano River. These facilities occupy a very small proportion of the drawing's width. In contrast to the two-dimensional depiction of these facilities, "Hiiragiya" is depicted three-dimensionally. Obviously, this was not the long-established *ryokan* (traditional inn), which stands to this day near Kyoto City Hall but was instead a branch establishment in Yase that was also known as "Warōtei."

From Yase Station to the Wakasa Highway, the illustration depicts a row of restaurants including Watariyanagiken (serving noodles, rice bowls, and sweets), Tagosaku, Benriken (selling *bentō* boxed lunches), and Eigetsu (specializing in beef *sukiyaki*). At the northernmost point, we find an area marked "passenger cars." Behind it, there is an expanse of paddy fields.

Let us now turn our attention to the left bank of the Takano River. We can clearly see restaurants, whether built using "a rustic style" remains to be seen, "dotted" in among an area dense with trees. Starting from the left, we see the following names: Heihachi Jaya; Shin Miura; Doi Shibazuke; Izumoya; Takomasa; Inari Jaya; Warabi Mochi; Hira no Imobō; Dōji Mochi; Darumaken; Suginoya; Uoshin; Sangenya; and Shiganoya.

The spot marked "power plant" refers to the Takano Hydroelectric Power Plant of the Kyoto Electric Light Company.

Interestingly, some of these are well-known establishments whose names may still be familiar to readers. "Heihachi Jaya" is a branch of Heihachi Jaya ("jaya" or "chaya" means teahouse in Japanese), which still has premises in the Yamabana area of Kyoto. The lines from Toyoko Yamasaki's *The Barren Zone* quoted at the beginning is a scene just after the main character Iki and his companions have gotten off the cable car and moved ahead "in the direction of a path along the Takano River..." As such, although the explanation that "This place is a roadside teahouse and inn that has been around since the Edo period" is incorrect, it is true of the main branch of Heihachi Jaya in Yamabana. The group of buildings seen on the left bank of the Takano River in Figure 0.1 is the branch of Heihachi Jaya. According to an advertisement placed in the *Kyoto Nichinichi Shimbun*, a local newspaper, the Yase Amusement Park branch opened on August 31, 1926.[12]

"Shin Miura" in a wooded area on the riverside is a branch of a chicken restaurant that was in the eastern Nijō Kawabata area at the time. The main branch has since moved to Kami-kiyamachi, where it remains to this day. Its branches in the Yase Amusement Park were sometimes called "Ryusuitei" or "Yanagi Jaya."

Next to the cable car, we find "Doi Shibazuke," which appears to have been something like a tea house, albeit a somewhat unusual one with a focus on *shibazuke* pickles. Doi Shibazuke originated in Ōhara, which is close to Yase. Beside the torii gate at the top of the slope, we can see Inari Jaya, which appears to also be a teahouse, as its name suggests. "Izumoya," located midway between the torii and the cable car station, is named "Takimi Jaya" on another map and has a newspaper advertisement for "Edo-mae eel."[13] Given that Izumoya was an eel restaurant with multiple establishments in Kyoto in the prewar era, it is possible that Takimi Jaya was a branch of Izumoya.

If we return to the riverside, we see "Takomasa" and "Warabi Mochi" next to one another. Based on the characters used in the names, Takamasa appears to have been an establishment selling *kantodaki*, which is Kantō-style *oden* (assorted ingredients simmered in a savory broth). Visitors could grab something savory there, before heading next door for a sweet at Warabi Mochi.

At the center of the picture, next to the pond, we find "Hira no Imobō," a branch of Imobō Hiranoya in Gion Maruyama. "*Imobō*" is a dish that is not generally known outside Kyoto. We will trace its history in Chapter 1.

"Dōji Mochi" and "Akioka Photography" are located on either side of the power station. A little way along the path from there, we find "Darumaken" (a restaurant specializing in *oraga soba*), with Suginoya (*udon* and *soba* noodles) behind it.

We pass by "Ōtakifu" (referred to by Hatsusaburō as "Ryūōgataki" in Figure 0.2), where there is still a waterfall with a modest flow of water and, atop the hill, we find "Uoshin" and "Sangenya." The former may have some connection to Uoshin in Nishijin.[14] The name "Sangenya" actually refers to a restaurant and *ryokan* located on the left bank of the Katsura River (Hōzu River) in Arashiyama, described as "a ryokan above the Togetsukyō Bridge on the north bank of the Ōi River, ... [which] has been famous since ancient times."[15] The one in the figure was in fact a branch of the Arashiyama Sankenya. Shiganoya, located on the right side of the illustration, was a restaurant for banquets.

Of course, "The Refreshing Bounty of Yase in Summer" is not an accurate map, but instead depicts the entire amusement park in a single view. This somewhat stylized form of representation may have led the artist to unconsciously highlight the intended design of the space. The right bank of the river is a zoned development area comprising a leisure space to the south of the railroad tracks and a shopping street in front of the station on the way to the Wakasa Highway. Despite its prime location at the foot of Mount Hiei, based on the elements of the landscape alone, this area feels bland and featureless.

The left bank of the river has also been *systematically developed* with adept use of the terrain at the foot of the mountain. The installation of a power plant by the Kyoto Electric Light Company is a feature not found in other amusement parks and has a significant impact on the geography of the park. Much of the water required to generate power is taken from the upper reaches of the Takano River and channeled through waterways along the elevation. Some of this water is diverted through an artificial waterway built on the slope, forming a waterfall, which is landscaped as a strolling garden. And we find the area "dotted with eateries constructed in a rustic style."

The original text from the official history of the Kyoto Electric Light Company states that the area was "dotted" and suggests that the restaurants were deliberately invited to establish premises in these locations. In fact, a contemporary account more directly states that "The Yase branch of Heihachi Jaya was opened in 1925, when the train line ran from Demachiyanagi to Yase, along with several other famous restaurants, at the request of the train [company]."[16] It appears that Zensuke Ōsawa played a central role in selecting and inviting famous restaurants from in and around the city.

Borrowing the name of the local gourmet magazine *Rakumi*, first published in 1935, we can say that Yase Amusement Park was a veritable paradise of "rakumi" (roughly meaning "flavors of Kyoto")[17] where the natural topography and scenery at the foot of the mountain were adeptly harnessed to create a park spread over a large area, and dotted with restaurants "in a rustic style" to create a distinctive culinary theme park.

4. The structure of this book

This book takes the restaurants and cuisines of the culinary theme park that was the Yase Amusement Park and uses this as a springboard for reexamining the culture of *rakumi* – "the flavor of Kyoto." Taking another look back at "The Refreshing Bounty of Yase in Summer" (Fig. 0.3) from the left, this section will provide an overview of the cuisine and restaurant culture covered by each chapter.

Heihachi Jaya was located a little upstream from the Saitōbashi Bridge. This establishment was equipped with the famous *kamaburo* (steam bath) of Yase (Fig. 0.4), but what set its cuisine apart from others? Let us quote an exchange from *The Barren Zone* again; this one between a waitress and the person who had invited Iki to the restaurant:

> ... A waitress dressed in the garb of an Ōhara woman came carrying refreshments and sake. She said, "What can we make for you today? The owner is eagerly awaiting your order in the kitchen."
>
> "I'd like to know if you have any *koi-no-arai*, salt-grilled sweetfish, or anything delicious from Wakasa," Kiji asked like a gourmand.
>
> "We just got a nice piece of tilefish."[18]

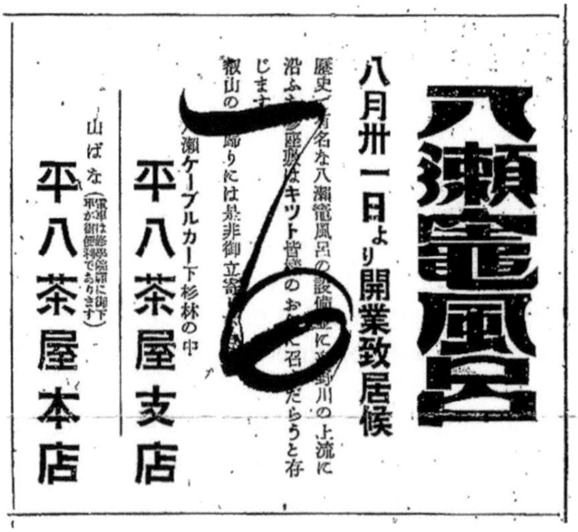

Figure 0.4 Advertisement for the opening of the Heihachi Jaya Shiten

In the original text, the word for "tilefish" is written with the phonetic characters for *amadai*, but it is likely that in her Kyoto dialect the waitress would have pronounced it as *guji*, another name for tilefish used in Kyoto and Osaka.

When lightly salted tilefish were brought in from Wakasa, the salt had soaked into the flesh just enough to make them perfect to eat when they reached Yamabana. The main branch of Heihachi Jaya served this with *tororo* (grated yam) soup and barley rice. The *mibu kyōgen* play "Yamabana Tororo" derives its name from this custom and informs us that "The transportation system of today has completely changed, so we rarely use seafood and serve mostly freshwater fish,"[19] echoing a phrase from *The Barren Zone*: "[the area] became known for freshwater fish dishes from around the time of the previous generation."

With its main restaurant and branches situated facing the Wakasa Highway and the Takano River, Heihachi Jaya must have originally excelled at both saltwater and freshwater fish dishes. In fact, the 1877 *List of Famous Places in Kyoto* states that "the Heihachi restaurant serves barley rice, *tororo* soup, *hisagi*, and even eel and carp, which will be especially delicious if suited to one's tastes." In the guidebook *Daikyōto* (roughly meaning Great Kyoto), published in 1928, Heihachi Jaya was classified as "general freshwater and saltwater fish cuisine."[20] This phrase "freshwater and saltwater fish cuisine," although not often heard today, was once the most common genre in the catering industry in modern Kyoto (Chapter 2).

Prominent politician Shinpei Gotō (1857–1929) reportedly said to Tenmin Matsuzaki (1878–1934), a journalist involved with a publication called *Shokudōraku* (The Gourmand) "If you go to Kyoto, eat the *suppon* [soft-shell turtle] from Ōichi rather than *mizudaki* [chicken hotpot] from Shin Miura."[21] Gotō's deliberate reference to Shin Miura while he was recommending Ōichi suggests that Shin Miura must have been well-known among gourmands. A branch of Shin Miura was situated downstream from Heihachi Jaya, across the bridge. In search of the origins of "*mizudaki*," for which Shin Miura has taken out a registered trademark, we will journey to the former brothel district of Yanagimachi in distant Hakata (Chapter 4).

We have already mentioned that the predecessor of Izumoya was Takimi Jaya, with its "Edo-mae eel." The term "Edo-mae eel" appears to suggest that the eel was grilled in the Edo (Tokyo) style, rather than that it was caught in the Tokyo Bay area. Following the history of the introduction of *Edo-yaki* into Kyoto will take us once again to the former brothel districts, with a brief historical and geographical excursion in Chapter 3 to the former Nezu brothel district in Tokyo.

Seichō Matsumoto's novel *The Globular Wilderness* includes a scene wherein the young protagonist (Kumiko) thinks, "I had come all the way to Kyoto, and I wanted something I couldn't eat in Tokyo." Then, after having a certain experience she "wants to eat Japanese food," and she remembers that she "had heard of a special dish in Kyoto called *imobō*," so she heads to a restaurant "inside

Maruyama Park" that the hotel receptionist had told her about: the main branch of Imobō Hiranoya.

> All the maids spoke the Kyoto dialect, and that was also the accent of the men in the adjoining room. Listening to the local language as she ate these unique dishes, she had a keen sense of having set out on a journey.[22]

Kumiko gains a true sense of the flavor of her journey "listening to the local language" and eating "unique dishes." We will discuss the *imobō* that Kumiko eats in Chapter 1.

There is another establishment in Yase Amusement Park that cannot be overlooked, although it's not mentioned in "The Refreshing Bounty of Yase in Summer." If we cross the suspension bridge to the east of the Hiiragiya branch and turn left before the waterfall, we see a long building off to one side. In Hatsusaburō Yoshida's "Bird's Eye View of Sightseeing Transportation for Famous Locations of Mount Hiei" (Fig. 0.2), this building is labeled as a "Western-style restaurant." Its name is Yase Shokudō, a branch restaurant of Manyōken, a pioneer in the world of Western-style dining in Kyoto. Chapter 5 will examine the spread of Western-style cuisine, which is closely related to the geisha districts. The final chapter will explore culinary sacred sites (within the precincts of a Shinto shrine) before bringing this book to a close with a short story that recently became known to the authors.

Notes

1. Yamasaki, Toyoko, *Fumō Chitai* (The Barren Zone), *vol. 3*, Shinchō Bunko, 2009, p. 22.
2. Nishimura, Zenshichirō, ed., *Daikyōto* (Great Kyoto), Daikyōtosha, 1928, p. 126.
3. Yanagita, Kunio, *Shufūchō* (Autumn Wind Diary), Sōgensha, 1940 [originally written in 1909], p. 154.
4. Osawa, Zensuke, *Kaiko 75 nen* (Retrospective75 years), Kyoto Hinode Shimbun Printing Department, 1929, pp. 234–235.
5. Kyoto Electric Light Co., Ltd., ed., *Kyoto dentō kabushiki gaisha 50 nen shi* (50-year history of Kyoto Electric Light Co., Ltd.), Kyoto Electric Light Co., Ltd., 1939, pp. 189–192.
6. Kyoto Electric Light Co., Ltd., *Kyoto dentō*, p. 190.
7. We will not touch on the transportation network on the Ōmi side here; we will discuss tours of Eizan on another occasion.
8. Nishimura, *Daikyōto*, p. 126.
9. Kyoto Electric Light Co., Ltd., *Kyoto dentō*, pp. 189–190.
10. *Yase-Ōhara no shiori* (The guidebook to Yase-Ōhara) edited by Kanezaburō Tanikita, first published on November 10, 1925 (Muchikuan, 1927, enlarged 3rd edition, p. 1), states that, "Since the railway opened on September 27, 1925, many have come to admire the scenic beauty and explore the historic sites of Yase-Ōhara. Every day, many visitors come and go. The Kyoto

Electric Light Company has also established a large amusement park here, which is steadily nearing completion." Additionally, the Eizan Cable Car and Ropeway website mentions the "opening of an amusement park" in the entry for December 1925. Eizan Cable Car and Ropeway (https://eizan.keifuku.co.jp/fun/history.php), last viewed on October 6, 2021.

11. Published in *Kyoto Nichinichi Shimbun* (June 27, July 11, 1929). In the June article, it was titled "Dark Green Yase Amusement Park." The types of business establishments mentioned hereafter are estimated based on advertisements published in the *Kyoto Nichinichi Shimbun*.
12. *Kyoto Nichinichi Shimbun*, September 13, 1926.
13. *Kyoto Nichinichi Shimbun*, April 3, 1927.
14. Nippon Telegraph and Telephone Public Corporation Kinki Telecommunications Bureau ed. *Kyoto-shi shokugyō betsu denwa bangō bo* (Kyoto city telephone directory by occupation) (Nippon Telegraph and Telephone Public Corporation Kinki Telecommunications Bureau, 1963, p. 629), advertisements for the Nishijin and Yase Uoshin are listed side by side.
15. Fūgetsu, Shōzaemon, *Kyoto meishō shashin chō* (Photo album of famous places in Kyoto), Fūgetsudō, 1910.
16. *Kyō no aji mise annai kettei ban: Kyoto aji no mise 506 – Zenshi ban sōshū hen* (The definitive guide to Kyoto's culinary establishments: Kyoto Aji no Mise 506 – All-city edition omnibus), Fujita Publishing, 1979, p. 292. Additionally, the owner of Bunnosuke Chaya, who had a store in front of Kodaiji Temple at the time, said: "Actually, around the time of the enthronement of the new Emperor, Han Tanaka (President of Kyoto Electric Light Company) told me to definitely open a store in Yase" (Osaka Mainichi Shimbun Kyoto Branch Office, ed., *Kyoto Shin Hyakkei*, Shinjidaisha, 1930, p. 86). It seems that invitations to businesses were still being made after the park had opened.
17. *Rakumi*, No. 1, 1935. *Rakumi* will be used as an important source throughout this book.
18. Yamasaki, *Fumō Chitai*, p. 25.
19. Usui, Kinosuke, *Kyoto mikaku sanpo* (A gustatory stroll through Kyoto), Shirakawa Shoin, 1962, p. 192.
20. Fukutomi, Seisui, *Kyoto meisho junran ki* (List of famous places in Kyoto), Murakami Kambei, 1877, pp. 29–30. Nishimura, *Daikyōto*, p. 74.
21. Matsuzaki, Tenmin, "Keihan tabe aru ki" (Osaka and Kyoto food diary) (*Santo tabe aru ki* (Food diary of three cities), Seibundō, 1932), pp. 1–128. Quote from page 20.
22. Matsumoto, Seichō, "Kyūkei no kōya" (The globular wilderness) (*Matsumoto Seichō zenshū 6: Kyūkei no kōya, Shi no eda* (The complete works of Seichō Matsumoto vol. 6: The globular wilderness: Branches of death), Bungeishunju, 1994 [first published in 1962]), pp. 3–300. The quotation from pp. 151, 153.

10.2478/9784991351600-008

Chapter 1
A Kyoto Specialty Called *Imobō*

1.1 Something you can't eat in Tokyo

1.1.1 A criminal sits alongside detectives

There is a striking scene in Seichō Matsumoto's short story *The Face*, set in 1950s Kyoto. The culprit and two detectives, who are investigating a case without knowing the culprit's face, end up sitting at the same table in a restaurant.

Under a pseudonym, the culprit has summoned a witness to the crime that he committed, and the witness has journeyed from Yawata in northern Kyushu to Kyoto. The detectives perceive that the person who summoned the witness is the criminal and accompany the witness. The scene depicts the sequence of events from each side, leading up to the detectives' chance encounter with the criminal without realizing his identity.[1]

The witness and the two detectives...

> One of the detectives took out his watch.
>
> "It's twelve o'clock. Let's get a bite to eat and go to the station," he said.
>
> "Good idea. *If we're going to eat the same thing, let's get something called* imobō, *a local specialty*," said the other.
>
> "*Imobō*? That'll be expensive."
>
> "Expensive is fine. I reckon we've already overrun our expenses for this trip. I don't know if I'll ever come to Kyoto again, so let's go eat anyway."
>
> Following this conversation, they went to a restaurant near Maruyama Park behind Gion.

The culprit / protagonist...

> I felt a little hungry. I wondered what to eat. *I had come to Kyoto, so I thought I should try something I can't eat in Tokyo.* So, I decided to try *imobō*.
>
> I got off the train in front of Yasaka Jinja Shrine and headed up toward Maruyama Park.

The detectives are thinking that they won't come to Kyoto again, and the criminal is thinking that he wants to purposefully try something he can't get in Tokyo. By chance, they both choose *imobō* for lunch and end up sharing a table at "a restaurant near

Maruyama Park." What kind of dish is this *imobō,* a Kyoto specialty that people from northern Kyushu expect to be "expensive" and that one cannot eat in Tokyo?

1.1.2 *Meoto-taki* – A dish of taro and dried cod

Imobō is neither mentioned in the expansive *Encyclopedia of Food and Drink* by Kishū Motoyama (1881–1958) nor is there any related information in *The Dictionary of the Osaka Dialect* edited by Shiyō Makimura (1898–1979), which includes a wide range of food-related topics.[2] However, in *The Illustrated Encyclopedia of Dietary Life in the Edo Period,* which has entries on every aspect related to food from the early modern period, there is a reference to "*imobō*" in the entry on "cod." This encyclopedia was written by Osamu Shinoda (1899–1978), a researcher of the history and folklore of food, including sushi, who had thorough knowledge of Kyoto's cultural history.

> Kyoto is famous for its *imobō* which refers to an *umani* [a braised dish] made with dried cod that has been softened by soaking it in water and then simmering it in hot water with taro or *yatsugashira* [a type of yam]. The dried cod used for this is called *bōdara,* dried cod that has been grated into three pieces and dried without adding salt. The dish was therefore named "*imobō*".[3]

Thus, *imobō* ["*imo*" meaning taro or yam and "*bō*" coming from "*bōdara*"] is a dish made by mixing yams and *bōdara* dried cod, written as 芋棒 ("yam stick") in kanji.

A work written by three women born in the Taisho Era (1912–1926) discusses Kyoto's *obanzai* (a meal of side dishes), providing a concise and clear description of *imobō* that includes seasonality and customs. Below are two passages from the beginning of the September chapter:

> First day of the month: On the first day of every month, we cook *nishin kobu* [herring kelp]. This month too, we aim to live austerely and frugally with this dish of the astringent minced herring and sliced kelp.
>
> Fifteenth day of the month: *Imobō.* Combine *ebi-imo* [shrimp-shaped taro] and *bōdara* dried cod in *meoto-taki* ["husband and wife dish" – a combination of deep-fried and grilled tofu]. At home, use ordinary taro buds.[4]

Because "the first day of the month is a day of renewal, care must be taken in preparing all things, even food."

> The fifteenth day is in the middle of the month, and it too is a day of renewal, like the first day of the month. That being so, for lunch, we cook *aka-gohan* [rice with azuki beans] and

> *imobō*, which is taro buds and *bōdara* dried cod cooked with *meoto-taki*, with *onamasu* [raw fish and vegetables in vinegar] as a side dish.
>
> ...
>
> When cooking the taro buds at home for the *imobō*, just use normal taro, not *ebi-imo* like they would use in a restaurant.[5]

Another source confirms that "eating boiled taro and *bōdara* dried cod was originally an old custom in the *machiya* townhouses of Kyoto, which was generally observed on the first and fifteenth days of the month," so it seems that it was a household dish that was associated with "days of renewal."[6] Apparently, it then became a custom that was observed only on the fifteenth day of the month.

Which means that Seichō Matsumoto's characters in *The Face* went out of their way to go to "a restaurant in Maruyama Park behind Gion" to eat *imobō*, a familiar home-style dish in Kyoto.

> At the center of the park is the famous *imobō* of Hiranoya. Even those who do not know the name of Hiranoya, will know it simply as "*imobō*."
>
> *Imobō* is the specialty of this place, a combination of taro and *bōdara* dried cod. In Kyoto, it has long been customary to eat *aka-han* (azuki bean rice) with *imobō* on the first and fifteenth of the month. It should be said that this is an ingenious culinary practice of Kyoto, where fresh fish is scarce.[7]

As we can see, according to Kinosuke Usui's *A Gustatory Stroll through Kyoto*, a guidebook on the food of Kyoto that was most influential in writing this book, Hiranoya in Maruyama Park is a restaurant which, even if you don't know its name, you will find your way to in search of the *imobō* that is its specialty.

1.2 The *otafuku-mame* of Hiranoya

1.2.1 A chrysanthemum field in Makuzugahara

Hiranoya features in virtually every guidebook to Kyoto written since the Meiji period (1868–1912). Guidebooks from the early Meiji period mention "Yasaka Shrine Kitabayashi Hiranoya" and "Gion Makuzugahara Hiranoya" in the section on *shōjin ryōri* (Buddhist vegetarian cuisine).[8] Kitabayashi is located between Yasaka Shrine and Chion-in Road, and to this day, it comprises a string of restaurants running east to west. Meanwhile, Makuzugahara refers to the area from present-day Maruyama Park to the Kikukei (Kikudani) River on the northern edge of the precincts of Kōdai-ji Temple, which has been a tasteful recreational area since the Edo period. Interestingly, the *Illustrated Guide to Famous Places of Kyoto* (1887) lists "Gion

Makuzugahara Hiranoya" under *shōjin ryōri* and "Hiranoya" under *sekigashi ryōri*, meaning places where geisha entertainment is available. Other establishments listed in this category include:[9]

Maruyama: Yaami Seiami Saami
Makuzugahara: Botanen
Maruyama: Fujinodana
Maruyama: Hiranoya
Sanbongi: Gepparō

Judging from the fact that Hiranoya is listed alongside establishments such as Yaami and Saami, which were sub-temples of Maruyama Anyōji Temple, and alongside Gepparō in the former entertainment district of Higashi Sanbongi, it appears that at that time, Hiranoya may have also offered *sekigashi* entertainment.[10]

Hiranoya is also listed in *Pioneers of the Capital*, a guide to commerce and industry published in 1883 (Fig. 1.1). On the mountains in the background of the illustration, the words "Kiyomizu," "Ōtani," "Shōgunzuka," "Onsen," and "Chion-in" are arranged from right to left (south to north), suggesting that this is a bird's-eye view from low in the sky to the northwest. The central part of the image is mostly occupied by expansive grounds labeled "Hiranoya," a two-story building with a large hall. Although this depiction may be somewhat exaggerated, it provides a picture of Hiranoya in the early Meiji period.

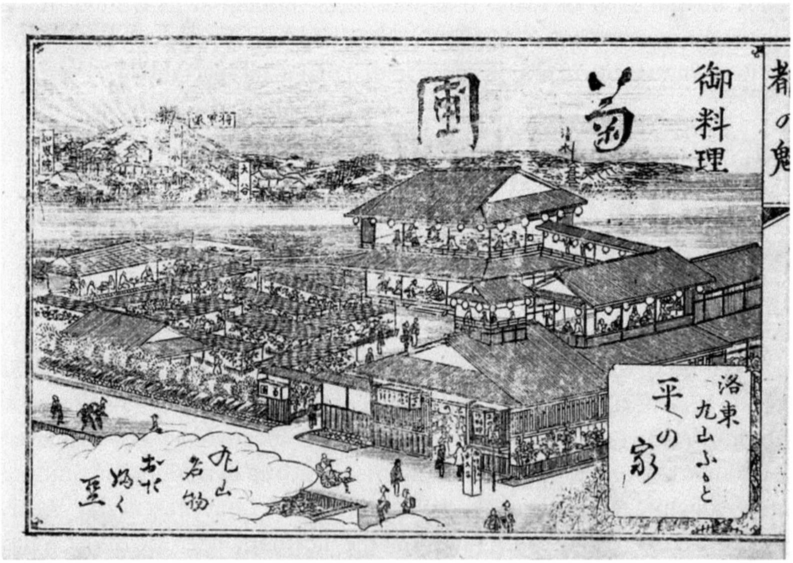

Figure 1.1 Hiranoya in the early Meiji period

Other than the enormity of the building, what immediately catches the eye in this drawing are the peculiarly large characters reading "Kikuho" or "Kikubatake" (菊圃 – "chrysanthemum field") next to those saying "cuisine" in the top right corner. We can deduce the meaning of this from the following quote from the *Guide to Famous Places of Kyoto* (1893):

> If one leaves Yasaka Shrine and heads east toward the Maruyama-michi, there will be a large weeping cherry tree on the way[.] When it is in full bloom in the season of spring flowers, its splendor attracts many visitors[.] When a bonfire is lit under it at night, it is truly splendid, as if one had been enveloped in a white cloud spanning all the heavens[.] This is called Gion no Yoisakura [Evening Cherry Blossoms of Gion][.] The left and right sides of the Maruyama-michi are called Makuzugahara, a place where poets have gazed at the moon and recited poetry since ancient times[.] *On the right is a restaurant called Hiranoya*[.] *This place is also called Kikuho [chrysanthemum field], and there are many rows of chrysanthemum flowers in the garden in autumn*[.] To the east is a forest of cherry blossoms[.] To the left is Botanen [lit. "peony garden"], where one should see peonies in the season of early summer[.] To the east of here is Maruyama Sōmen, the main gate[.][11]

Thus, we see that Hiranoya was a restaurant that was referred to as "Kikuho" or Kikubatake," with its autumn display of chrysanthemums in the garden. Indeed, on closer inspection, we see that the entrance to the garden in Figure 1.1 has these very characters, 菊圃 (Kikuho), written on a signboard, and the garden is shaded from the sun, creating a space where guests could appreciate the chrysanthemums in boastful bloom in the flower beds.[12]

At that time, Hiranoya faced Maruyama-michi, which ran from Yasaka Shrine to Maruyama Sōmon Gate to the east.[13] The road depicted in the foreground of the image from Figure 1.1 is probably Maruyama-michi.

We get a glimpse of the dishes that were served at Hiranoya in the Meiji period from an article called "The *Imobō* of Hiranoya" by Doppo Kunikida (1871–1908), a journalist and novelist:

> If one were to review one of the most famous and inexpensive cuisines of this area, it would be the branch of Hiranoya in Maruyama Park. Though this is a modest establishment, it is by no means of an undignified character to stop by for a while. It serves various dishes, rice with beans in late spring, rice with chestnuts in the autumn, and what is called "umewan" ["plum bowl"], a bowl with 45 different items, including fish cakes, tofu skin, shiitake, and so on. Then, we have *tōkake tofu* (albeit in a small bowl) and a plate of *imobō* (softened salted cod simmered in a sweet broth with *yatsugashira* yam), thus a meal costs only 25 *sen*. The value is very great compared to the price, and this establishment flourishes greatly.
>
> Of all the aforementioned items, the most prized by the locals is *imobō*, which is made using taro that are extremely large yet soft and of good flavor and quite unlike

> anything else. However, as this is a lower-class dish, it is not suited to upper-class tastes. The name *imobō* is an abbreviation of *imo* and *bōdara*. This establishment also has the famous *otafuku-mame*, referring to a dish of sweetened *soramame* [broad beans] cooked to retain their shape, that one should use in place of *kinton* [mashed taro with sweetened chestnuts] as a sweetener. More than a few people bring them back to Tokyo as souvenirs during the cold season, and one can order these beans for merely 25 *sen* or so.[14]

Here, Doppo explains that "the *imobō* of Hiranoya" was a famous Kyoto dish that was also inexpensive. He evaluates the flavor very highly, but also points out that "this is a lower-class dish, ... not suited to upper-class tastes" (the authors agree). We will later discuss the fact that this was a "branch" restaurant.

For the moment, note the dish in the last paragraph, *otafuku-mame*, described as being a popular souvenir to take back to Tokyo. In the lower left corner of Figure 1.1, we do indeed find "Maruyama's famous *otafuku-mame*." Curiously, Figure 1.1 contains no reference to *imobō*, although the curtain in the entryway reads "Hiranoya" and the paper lantern says "cuisine." Instead, we find "*otafuku-mame*" written in large letters as a "specialty." Both *imobō* and *otafuku-mame* may have been signature dishes of Hiranoya.

In fact, Tokujirō Honda, the owner of a restaurant in Tokyo, discussing the "characteristics of Kyoto cuisine" in a women's magazine, explained: "*imobō* is a dish made by simmering *ebi-imo* and *bōdara*... This is known to people in Kyoto as a family tradition of Hiranoya ... *Otafuku-mame* are also a family tradition of Hiranoya, made using *soramame* broad beans."[15]

There are also frequent references to "*otafuku-mame*" in relation to the main Hiranoya branch in Kyoto, although it has been described as being known "simply as '*imobō*.'" For example, local historian Ryokkō Tanaka (1891–1969) notes that "below [the mineral spring] there is a peony field, and fields where *otafuku-mame* grow well,"[16] also noting that Makuzugahara, the precursor to Maruyama Park, was farmland. Thus, like the name "Kikuho" (chrysanthemum field), "Botanen" (peony garden) was listed as a *sekigashi ryōri* establishment and may have been a restaurant where guests could view peonies. Tokujirō Honda, mentioned above, explains that "The beans were once produced in Maruyama, yet the eastern part of this area has become too open and the ground to plant beans and so on has been lost, so we are now seeking a region called Aza-ama near Settsu Sumiyoshi where they can be grown."[17] Tsuneji Ōkubo (1897–1983), an Osaka-born food critic, recalls:

> When I was a child, which is some 30 years ago now, there was a truly delicious dish of simmered beans made from broad beans that had a tortoiseshell color. Though this is now an old memory, I cannot forget that color and shape and how delicious they were. When I met a business proprietor, Mr. Kitamura Fujinosuke, and told him of this memory, he said, "I would like to try simmering those tortoiseshell beans. But the first question is whether they would be like they were in the old days, and besides, unlike when one is

doing it at home, one would have to boil up a considerable amount for one pot, and they won't keep for very long, even in winter. It would be a waste if there were any leftovers, so I haven't tried it yet."

All the same, I am begging here to eat those simmered beans at least once more.[18]

Unfortunately, it seems that there never was a revival of *otafuku-mame*.

1.2.2 A secret family recipe

Although one may find themselves saying that *"imobō" is nothing more than a home-style dish of simmered ebi-imo and bōdara*, and that may indeed be so, the reason that people are drawn to this homely dish, similar as it is to a side dish, is perhaps because of the high quality of the taro and *darabō* used therein, but also because there is a knack to how the dish is simmered. Saying there is a knack may make it seem that this is a difficult dish to prepare, yet this is not the case. In fact, what makes the dish attractive is the kindly manner, so to speak, in which the cook sets out to prepare the dish, as it instills feelings of familiarity in anyone [who eats the dish].[19]

Although "*Imobō* is nothing more than a home-style dish," food connoisseurs from both Kyoto and farther afield have been drawn to the *imobō* of Hiranoya.

Let us now turn our attention to the "Commentary on Imobō" by Tōnosuke Kitamura (1892–??). He became head of the family behind Hiranoya in the early Shōwa era (mid-1920s). In 1928, in anticipation of the enthronement of the Shōwa Emperor (Emperor Hirohito), Hiranoya published the "Guide to Excursions from Higashiyama to Eizan, Focusing on Hiranoya's Famous Kyoto Dish of Imobō"[20] in which Kitamura provides a detailed commentary on the history of Hiranoya and *imobō*. Admittedly, the fact that it is self-promotion strongly colors the text as an advertisement for the restaurant; but it is nonetheless a valuable source and is worth citing extensively.

... *imobō* is welcomed as a famous dish of Kyoto, not because the curtains at the entryway to our restaurant are long-standing, but rather because its price makes it feel familiar. And yet it is not because it is inexpensive, nor is it because of the long history attached to *imobō*. It is, in a word, all about the taste. The skill of cooking lies in not losing the original quality of the ingredients, while at the same time, the peerless art of cooking lies in making something that is not naturally delicious into something delicious and serving it forth.

The terms *imobō* and especially *bōdara* are used in Kyoto to denounce someone who lacks wisdom, and indeed, being almost tasteless and bland, they are not as delicious as the sea bream or the pike conger eel; the seasoning must be correct in order to draw out the flavor. The value of this famous dish of Kyoto lies in the combination of taro and this *bōdara*, and the use of a unique seasoning method that makes its flavor even better than that of sea bream, creating a clever dish worthy of setting before our customers.

> Once one places a piece of taro on the tongue, it seems to melt, and when the *bōdara* has passed one's lips, one experiences its softness and a flavor beyond words. Moreover, our restaurant has long been praised by gourmands, who mention that the seasoning methods are appropriate for the region of Kyoto, among other things. I would desire for all to taste for themselves whether this is true or not, and to know the efforts our ancestors went to, and at the same time, to criticize this dish as a Kyoto specialty and compare it with specialties from around the country.

Thus, Kitamura believes that the reason why *imobō* is recognized as a specialty is not because of its history or its reputation but because of the "taste" itself. He also believes that the essence of cooking is to make "something that is not naturally delicious into something delicious" and that the pride of a restaurant lies in doing this while respecting the ancestors.

Kitamura goes on to explain that in Kyoto, where fresh saltwater fish was hard to obtain, salted and dried fish had long been treasured, and especially the combination of *bōdara* and taro had come to be "appreciated as an unparalleled side dish of excellence in Kyoto." *Imobō* came to be recognized as a "delicacy of the mountains and seas" and even as "auspicious of marital harmony." The "traditional custom" of eating *imobō* on the first and fifteenth of each month, as well as on New Year's Day, continued to be practiced until at least the Second World War. Kitamura also explains that a daughter-in-law "could become well liked by her mother-in-law by simmering an impeccable *imobō*."

Hiranoya and its *imobō* came into being through overlapping coincidences in the geography and history of Kyoto.

> The ancestor of our family was Hirano Gondayū, and he served Prince Awata Shōren'in during the An'ei era, charged with the responsibility of cultivating chrysanthemum fields [*kikuho*] for the imperial household, at the same time as growing vegetables, and receiving orders from the Imperial Palace. Moreover, as a renowned gourmand, his knowledge of food was very great, and his cultivation methods in particular have been deemed without equal both then and now.
>
> One year, Prince Shōren'in *had taro seeds brought back from Nagasaki in Kyushu* and had my ancestors cultivate them; they were eminently suited to the local soil, and produced a large crop of a quality that is unparalleled in any era. The prince was so delighted that he named it *ebi-imo*, and greatly praised it. This is the origin of the currently renowned *ebi-imo*. This high-quality taro was immediately associated with *bōdara* in Kyoto, and as a matter of course this was then further investigated in Kyoto, and was quite natural that our ancestors, who were both food connoisseurs and culinary investigators, took great pains to learn how to cook these ingredients. It is of the utmost difficulty to season the dish so as to find a balance of hardness and softness – the *bōdara* being hard when simmered, and the taro which softens immediately when simmered, with their skins as soft as fine silk – and to bring about mutual harmony in their tastes.

The founder of Hiranoya (lit. "house of Hirano") was Hirano Gondayū, who served Prince Shōren'in during the An'ei era (1772–1781). He was adept at managing chrysanthemum fields and cultivated vegetables for the Imperial Palace and was known as a gourmand. As we can see, seed taro were brought to Gondayū from distant Nagasaki. By happenstance, they were well suited to the soil of Makuzugahara. This is the origin of the *ebi-imo*, "shrimp taro," which is so-called because its curved shape is thought to resemble a shrimp. Further, "this high-quality taro was immediately associated with *bōdara* in Kyoto." Gondayū, prevailed in the "utmost difficulty" of preparing these ingredients, was recognized by Prince Shōren'in under the trade name of "Hiranoya" and started a culinary business.

> Thanks to this, the name *imobō* is now known far and wide across the land, and one may say that it has acquired the status of an unwritten rule for sightseers, that if one once sets foot in Kyoto, one must surely enquire after the paradise of Maruyama, and if one visits Maruyama, one must surely taste *imobō*.

All gourmands in the 1950s and 1960s who were familiar with Kyoto loved Hiranoya's *imobō*. However, they were also aware that "this dish called 'imobō' is by no means a rare dish but rather an ordinary one,"[21] or that "the *imobō* of Hiranoya is a side dish for households in the Kyoto area, eaten not just at New Year's, and is not a high-class dish made by experts."[22] There was also a perception that "as it is a side dish that any household can make, Hiranoya wouldn't be able to attract customers if it weren't particularly skilled at making the dish" and that "unlike other dishes, there is no secret or trick to it, and people buy it for its genuine value, so cheating won't work."[23] It was therefore necessary for Hiranoya to continue offering a version of *imobō* with a flavor different from that of ordinary households, based on the *ebi-imo* and cooking methods that originated with Hirano Gondayū.[24]

"The seasoning method has been passed down from one son to the next, and for 170 years, we have dedicated ourselves to our primary business with a single-minded purpose." This is referring to the concept of *isshi-sōden*, whereby the closely guarded secret recipe has been handed down so that it is known only to one member of each generation of the family, which is perhaps the reason why *imobō* has become a specialty of Kyoto.

1.3 Genealogy and place

1.3.1 The location of phone number 155

> It is said to have moved frequently from one place to another in this area before finding this location...[25]

> Its location within the park seems to have changed frequently, and yet it has already been a long time since it settled in its current place.[26]

There are currently two branches of Hiranoya in the Maruyama Park area: Imobō Hiranoya Honten and Imobō Hiranoya Honke.[27] During the Meiji, Taishō, and prewar part of the Shōwa period, Hiranoya opened multiple restaurants. In this section, we will trace the locations of their establishments with a focus on Gion Maruyama, the main area of their expansion. The tools used to trace the genealogy of their restaurants and locations are a list of merchants and businesses called *A Record of the Names of Persons in Commerce and Industry*, and the telephone numbers listed in advertisements.

First, in 1905, we have "Maruyama-chō, Hiranoya: Kitamura Kumezō" (phone number 155) and "Maruyama-chō, Imobō Shiten: Kitamura Kumezō" (no phone number) listed alongside each other (the proprietor's name is listed after the colon in each entry, and the same applies below).[28] Thus, at this stage, Hiranoya was operating at least a main restaurant and a branch restaurant (*shiten* in Japanese) using the trade name "Imobō" in "Maruyama."

In 1907, we find "in front of the Gion torii gate, Tabiyado Hiranoya: Kitamura Kumezō" (phone number 155) and "Gion Maruyama-chō, Imobō Shiten: Kitamura Kumezō" (no phone number).[29] In fact, Yaami and other ryokan and restaurants in Maruyama Park and the surrounding area, were destroyed by a succession of fires throughout the Meiji period and on April 7, 1906, in the midst of bustling crowds of people who had gathered to catch a glimpse of the weeping cherry tree, Hiranoya, too, was burned to the ground by a careless customer's cigarette.[30]

It was the main restaurant that was destroyed by fire in 1906. In 1907, "in front of the Gion torii gate, Tabiyado Hiranoya: Kitamura Kumezō" opened its doors for the first time, but the old branch restaurant still existed, so we can deduce that the main restaurant had moved outside Maruyama Park. In fact, the new restaurant also inherited the 155 phone number, and there is an interesting point connected to this.

> I do not believe that there are many people who know that there was a Hiranoya located near Shinkōmichi in Shimogawara. Here, *a restaurant was opened that was separate from the* imobō *of Maruyama*, and it was once a busy establishment, along with Nakamurarō in front of the torii gate.[31]

This restaurant was located close to Shinkōmichi, which is on the south side of the precincts of Yasaka Shrine. Given its location "in front of the Gion torii gate," we may assume that it was located at the northern end of Shimogawara-dōri, immediately outside the torii gate. In fact, the following explanation is offered: "Hiranoya, renowned for its *imobō* and *otafuku*, was burned out of its location in Maruyama at the corner in front of the torii gate two years ago; it then took over from Toganoo and has been prospering and cultivating enviable chrysanthemums."[32] Thus, Hiranoya took over the premises of Toganoo, which had been an iconic restaurant of the early Meiji period.

The important point here is that the main branch of Hiranoya did not use "*imobō*" in the name of the restaurant or its trade name. If we cast our minds back to what Doppo Kunikida introduced as "one of the most famous and inexpensive cuisines of this area," which he then described as "the branch of Hiranoya in Maruyama Park," it becomes clear that Doppo was enjoying *imobō* at a branch restaurant.

We can now establish a hypothesis: The Hiranoya that was founded by Hirano Gondayū did not necessarily specialize in *imobō*. Although it would have served *imobō*, it must have also served other dishes. However, in the latter half of the Meiji period, a specialist branch restaurant was opened, featuring the word *imobō* on its signboard.

In a register of businesses from 1905, we find the following:

Gion Maruyama, Imobō: Kitamura Kaichirō (phone number 1603)
Gion Maruyama, Hiranoya: Kitamura Kumezō (phone number 155)
Gion-chō, south side, Hiranoya Shiten: Kitamura Teru (phone number 2770)

In this way, we have the words "Hiranoya" and "Imobō" listed alongside one another, while another branch restaurant has emerged in the southern part of Gion.[33] It appears that the main restaurant had moved back to Gion Maruyama from its location in front of Yasaka Shrine's torii gate. The address of the branch on the south side of Gion changed from "Gion-cho, Hanamikōji" to "Shijō Yamato-ōji Higashi, Imobō Hiranoya Shiten: Kitamura Fujinosuke" in 1924.[34] Fujinosuke, whose name appears here, quickly made his mark, and later became involved in the construction of Yamakage Shrine in Yoshidayama (which we will discuss in the final chapter), becoming a well-known face of the culinary world of Kyoto.[35]

If we jump back in time from this point for a moment, in 1920, the two Hiranoya establishments located in Maruyama Park moved about considerably. The person registered to the 155 phone number was changed to Kitamura Kaichirō, while the 1603 number changed to Kitamura Kumezō. These changes do not reflect a change of ownership, but a change in management between the main restaurant and the branch restaurant. Later, in 1904, the 155 number is not listed, and the 1603 number is once again registered to Kaichirō.

Then, in 1925, Kaichirō moved to managing the main branch of Hiranoya (No. 155), which is also in Maruyama Park, and Fujinosuke took over the management of the Imobō Hiranoya. The main restaurant was once again destroyed by fire in 1928.[36] The following year, on August 15, 1929, Fujinosuke opened a new establishment by the wisteria trellis in the park, using the words "Kaiseki Ryōri [Haute Cuisine], Hiranoya *Bettei*" (Fig. 1.2).

In the advertisement brochure *Guide to Famous Places of Higashiyama* ("Edition for the Main Restaurant of Hiranoya, Renowned Imobō of Kyoto Maruyama Park"), which was issued to mark the relaunch of Hiranoya, we find the address "Maruyama Park, in front of Chion-in South Gate, Imobō Hiranoya Kōten," listed alongside "Maruyama Park, Hiranoya Bettei." It is worth noting the phone number and description of the latter "Bettei" restaurant (meaning "villa") on the map. The "Kōten" (meaning "long-established restaurant") in front of Chion-in Temple has the numbers 1603 and 1604, whereas the "Bettei" has the number 2770, in addition to the previously existing 155. In summary, the Hiranoya management withdrew from the branch on Shijō-dōri – named "Hiranoya Shokudō," which will be discussed later – and transferred its phone numbers to the Hiranoya Bettei, which is marked as "Hiranoya Higashiten" ("eastern restaurant") on maps (Figs. 1.3 and 1.4).

Figure 1.2 Advertisement for the opening of Hiranoya Bettei

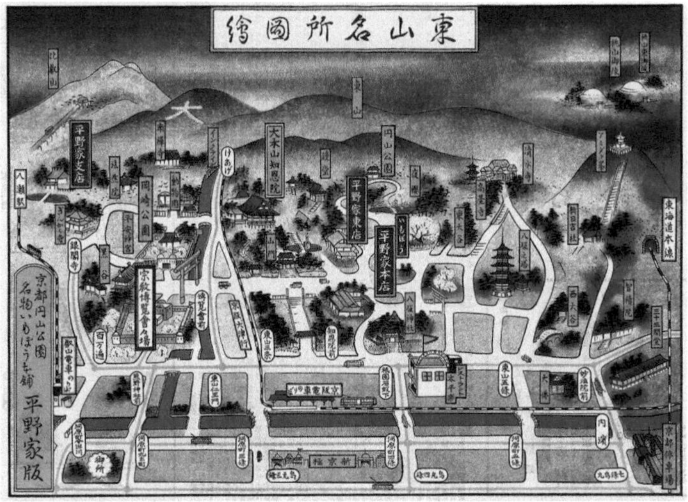

Figure 1.3 Hiranoya Higashiten depicted in an illustration

The meaning of the above should be clear: in short, this was the inception of "Imobō Hiranoya." The main restaurant had switched from being positioned as a restaurant to specifically being positioned as an *imobō* restaurant. Although it had been said that "the *imobō* [of Hiranoya] has been promoted as a Kyoto specialty for quite some time,"[37] the history of Hiranoya as illustrated above shows that relocating the main restaurant served as an opportunity for Hiranoya to make *imobō* their star attraction. It is no exaggeration to say that the "Kyoto specialty imobō" was created, at least in terms of the phrase itself, by Kitamura Fujinosuke.

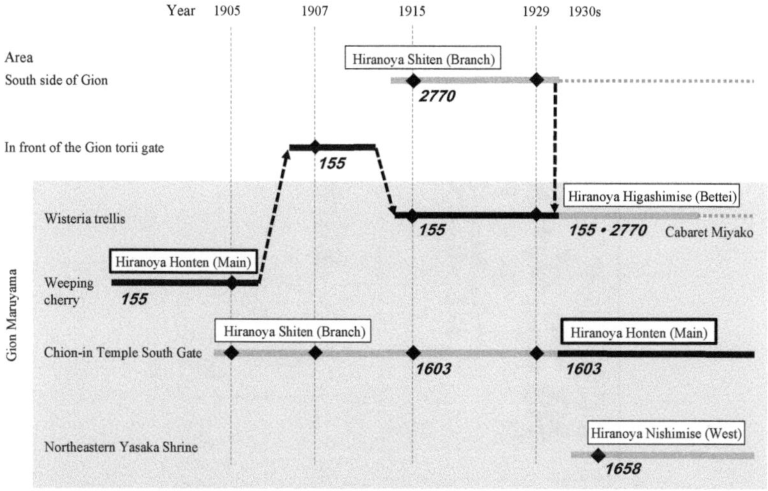

Figure 1.4 A genealogy of telephone numbers

The menu for each restaurant is provided in the store introduction of the *Guide to Famous Places of Higashiyama* as follows:

Kyoto Specialty – Imobō
Imobō, umewan, Gion tofu, rice, and pickles
90 *sen* per customer
In front of the southern gate of Chion-in Temple, Maruyama Park, Kyoto
Hiranoya Kōten ["long-established restaurant"]
Rakuhoku Yase Amusement Park, Hiranoya Shitei [Branch]

Kaiseki cuisine [Haute Cuisine]
Yuki menu (per guest): 1.50
Tsuki menu: 2.00
Hana menu: 2.50
Gion bentō (with soup): 0.80
Maruyama Park, Kyoto
Hiranoya Bettei

Thus, we can see that Hiranoya Bettei had taken over the role of the main restaurant before it burned down (Fig. 1.5).

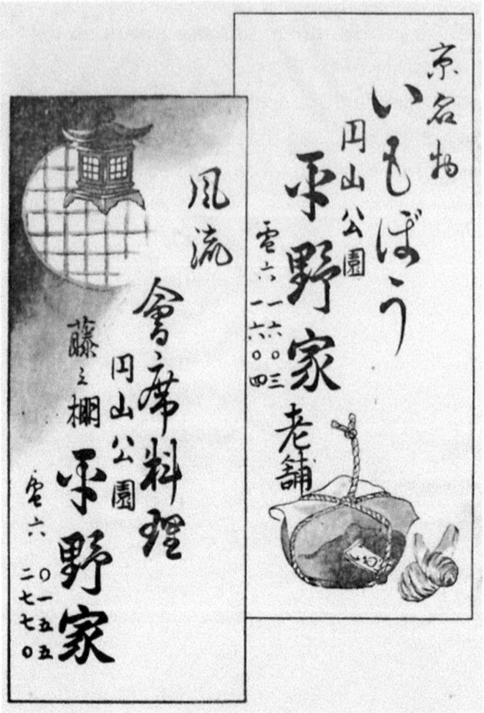

Figure 1.5 The *imobō* and kaiseki cuisine of Hiranoya

> Dishes from the Tsuki, Yuki, and Hana menus were served on the wisteria trellis and at the banquet in the 100-*tatami* mat hall, and Kimi, the young bride, served the guests with beautifully prepared sea bream and shrimp served on an *ōshima* ["large island"] platform in the center. The lights were turned off so that only Kimi was illuminated. It was a cutting-edge spectacle, and Kimi's glittering performance will be remembered.[38]

The Kimi mentioned in the quote was Kimi Kitamura, who married Fujinosuke at the end of the Meiji era. It seems that considerable efforts were made in managing the newly built Hiranoya Bettei to a similar degree as Hiranoya Kōten, which was the new main restaurant.

In this way, although the telephone number was first assigned to the Bettei restaurant or "east branch," a "west branch" also opened in the mid-1930s (Fig. 1.6), near the torii gate to the east of Yasaka Shrine. Although it was just around the corner from the main restaurant, this small geographical difference ended up creating a significant distinction between the two restaurants. Its proximity to the hall of Yasaka Shrine, though slight, meant that the western restaurant was subject to "forced evacuation" during the war. After the war, it eventually reopened as "Hiranoya Saryō" in 1957,[39] before later reverting to the name "Hiranoya Nishiten." This is the present-day "Imobō Hiranoya Honke" (Fig. 1.7).

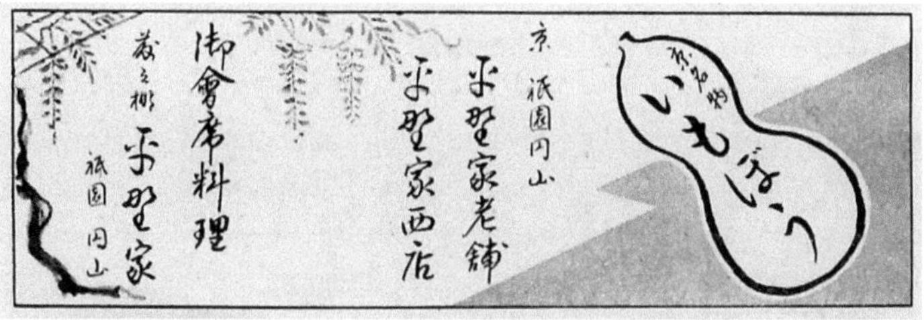

Figure 1.6 The three Hiranoya restaurants in Gion Maruyama

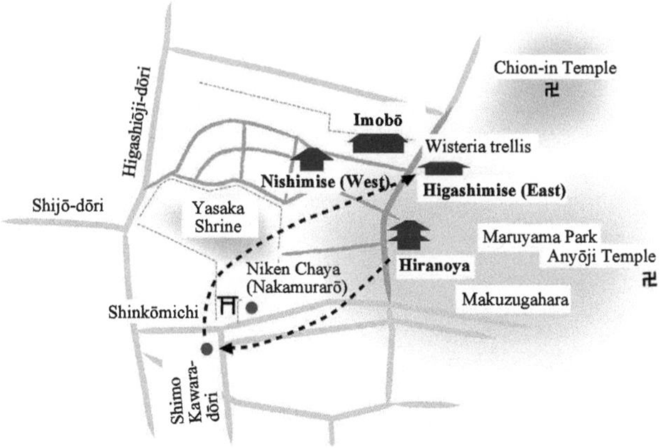

Figure 1.7 Locations of the Hiranoya restaurants in Gion Maruyama

Meanwhile, the fortunes of the eastern Bettei (Higashiten) by the wisteria trellis restaurant were rather more uneven: "After the war, it was requisitioned by the GHQ and used as a dance hall," and was "once again destroyed by an accidental fire."[40] Take a look at the advertisement in Figure 1.8. Its significance might be unclear just by looking at the advertisement for "Cabaret Miyako," but the location is given as "the wisteria trellis, Maruyama, Kyoto" and its phone number is (155).

This was the final usage of the 155 number, with its extensive pedigree in Gion Maruyama.

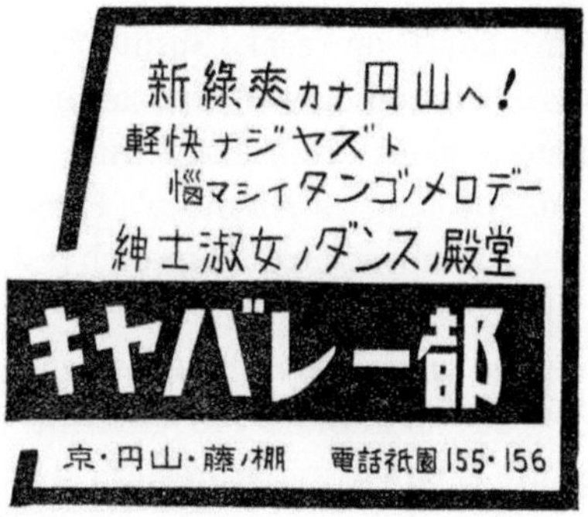

Figure 1.8 Advertisement for Cabaret Miyako

1.3.2 The deference of Hatsusaburō Yoshida

The "Guide to Excursions from Higashiyama to Eizan, Focusing on Hiranoya's Famous Kyoto Dish of Imobō" is a print featuring a bird's eye view created by Hatsusaburō Yoshida. The print is a view of the Higashiyama area from the west and centered on the main Hiranoya restaurant in Maruyama Park, which is the long, imposing building located in the middle of the image, with landmarks of Higashiyama, such as "Shōgunzuka" and "Chion-in Temple" dotted elsewhere (Fig. 1.9). The print presents a joyful image of spring, lush with the green of the mountains and the cherry blossoms in full bloom.

Figure 1.9 Main restaurant and branch of Hiranoya (Hiranoya Shokudō) drawn by Hatsusaburō Yoshida

Yase Amusement Park (Fig. 1.10) can be seen with an abundance of cherry blossoms. At the center is a prominently large farmhouse-style building with a thatched roof. Next to this is a building with open sides and a scarlet carpet on the floor. On the roof of the tatami-floored building overhanging the pond, there is a large sign that reads "Imobō," with the characters written from right to left. This is the "Imobō Hiranoya Shiten" branch restaurant in the Yase Amusement Park (see also the introduction to this book).

If we return to the right-hand side of the image, there stands an eye-catching modern white building in front of "Gion Yasaka Shrine." This is Hiranoya Shokudō, a branch of Hiranoya that was located on the southern side of Gion, in the "Ishidanshita" area to the south of the stone steps of Yasaka Shrine. The "Guide to Excursions from Higashiyama to Eizan, Focusing on Hiranoya's Famous Kyoto Dish of Imobō" advertises the features of the Hiranoya Shokudō branch restaurant as follows:

- This restaurant is designed to suit modern tastes, with a focus on convenience, hygiene, and economy – a restaurant for you [customers], not for us [owners].
- Although the architecture is purely in the Western style, the facilities inside are an ideal blend of Japanese and Western styles, where the shortcomings of each have been balanced out by the strengths of the other, to suit Japanese cuisine.

- The rooftop terrace offers superb views of Higashiyama, a panoramic vision of the entire great city of Kyoto.
- On the upper floors of the restaurant, each dining table has an electric light that creates a soft and warm ambience, making one feel right at home.
- Then, of course, there is the food[.] We take great efforts to provide our customers with seasonally appropriate dishes that are most suited to their tastes.

Figure 1.10 Branch restaurant of Hiranoya (Hiranoya Shiten) drawn by Hatsusaburō Yoshida

Regarding the external appearance of Hiranoya Shokudō, we have the following description:

> In the wind blowing down from the mountains of Higashiyama, flags flutter high upon the resplendent red and green two-storied gate of Gion's Yasaka Shrine – just to the left of Shijō Daibō, in a straight line from the *ishidan* stone steps of the shrine, is situated our Hiranoya Shokudō, which stands in splendid isolation, aloof yet charming, constructed in the early modern French style, it represents one aspect of modern Kyoto, while also embracing a quiet, calm, and homely elegance.

Hatsusaburō also wrote the following, entitled "A few words to accompany the picture:"

> In contrast to the main restaurant in Maruyama, which has an atmosphere of quietly faded splendor, Hiranoya Shokudo in Shijo is very bright and has a modern style that reflects an aspect of modern Kyoto. Yet as expected, it also has a sense of comfort where one can directly feel something like the dignity of a long-established restaurant that is remembered in some way. In addition, the Yase branch has a raised pavilion located on

> the banks of an elegant pond, where one can feel nostalgic over the rusty tones of the amusement park, which has recently settled into itself, and the scent of the imposing old cedars of overlooking Eizan – it is the perfect place to refresh oneself for the day.

The guide also presents photos of the main and branch restaurants. Comparing these with the illustrations, one wonders if Hatsusaburō's first-class drawing style is perhaps showing just a touch too much deference to his client, Fujinosuke. Nonetheless, the illustrations enable us to see (in an enjoyable manner) the development of Hiranoya in the early Shōwa period at a glance, which would elude us if we were just to trace the phone numbers over time.

Finally, let us read the impressions of Tenmin Matsuzaki, a master of touring from one eatery to another:

> Hiranoya of Maruyama serves Gion bento and lunch in a new café-style building in the Ishidanshita area of Gion below the stone shrine steps, but the best place to experience the Hiranoya feeling is in Nawanoren in Maruyama. *Imobō*, tofu dishes, and simmered beans – all had Hiranoya's special characteristic of being light, clean, elegant, relaxed, delicious, and reasonably priced. For about two yen per person, one can eat and drink elegantly – the softness of the taro, the seasoning of the cod – in these and other things somewhere lies the flavor of Kyoto: an old scent with new colors.[41]

Tenmin experienced this "flavor of Kyoto" in a Hiranoya in Gion Maruyama, rather than in the café-style Hiranoya Shokudō. He refers to this restaurant as "Nawanoren," referring to traditional curtains made of rope outside the entrance to an establishment. Rather than referring to the main restaurant, this was likely a branch restaurant at that time, which later became "Imobō Hiranoya Kōten."

1.3.3 Two views on the "flavor of Kyoto"

> The current generation said, "Well, we're not making much money, so I suppose we couldn't make ends meet without doing another job on the side at the moment. It just suited everyone's tastes, so this is how things ended up after getting the Hirano name from the nobility." It wasn't planned as a restaurant from the beginning. Spontaneously, so to speak, it became a house [or family] that cooked.[42]

The "current generation" refers to Fujinosuke, who overhauled Hiranoya's management systems in the early Shōwa era (late 1920s). What caught the interest of the interviewer (Shirō Takagi) was Fujinosuke's description of the spontaneous development of Hiranoya. Kansei Dōmoto (1887–1964) claimed that "the ancestors of the family running Hiranoya came to Kyoto from Yamato following Prince Shōren'in and opened a teahouse while cultivating taro fields in Maruyama."[43] Tsuneji Ōkubo also states:

> Hiranoya has the same origin story as Hyōtei at Nanzenji Temple.
>
> Just as Hyōtei began as a teahouse in front of Nanzenji Temple, [Hiranoya] *was a teahouse* in front of Chion-in Temple.
>
> Hiranoya, too, did not open as a restaurant; rather, it started as an establishment serving light refreshment to people coming and going to the temple. It followed the tastes of its customers, serving this and making that, before it became an establishment specializing in *imobō* as it is today. However, I am glad that, like Hyōtei, it has not forgotten its origins and has not become a business in the sense of a restaurant.[44]

We need not be concerned with whether it is true that Hiranoya originated as a teahouse. Instead, it is interesting that three critics who were all steeped in the food culture of Kyoto share similar impressions.

Originally, as we have seen, *imobō* was a dish that featured on the family table on the first and fifteenth of the month. The gourmand Jinnosuke Inoue (1905–1973), a critic in the Kyoto entertainment world who was also involved in the kimono industry, stated that "the *imobō* of Hiranoya, carefully and thoughtfully simmered, can truly bring out the flavor of simple, unadorned, home-style cooking." He goes on to say, "one can rightfully say that that '*imobō*' is truly the flavor of Kyoto, an unpretentious flavor liked by all."[45] In this way, *imobō*, the flavor of Kyoto, stands at a crossroads between home cooking and professional cooking. Sen Sōsa, the head of the Omotesenke tea ceremony school (at that time), similarly praised *imobō* as being "the true flavor of Kyoto, using the dried *bōdara* to bring out the flavor of the taro, making both excellent."[46]

Perhaps in response to the phrase "truly the flavor of Kyoto," Kitamura Fujinosuke fondly states that *imobō* is "a thing of encounters."[47] This is indeed true – an encounter between *ebi-imo*, an improved version of Chinese taro that was brought from Nagasaki, and *bōdara* brought from Hokkaido through the Sea of Japan.

Let us now return to *The Face* by Seichō Matsumoto. The culprit and the detectives unintentionally share a table in a tatami room at a restaurant that is likely Hiranoya (although this is not explicitly stated) to eat *imobō*, a specialty of Kyoto. Although it is a one-sided encounter from the culprit's perspective, this, too, is a form of encounter such as those mentioned above. The culprit chose to travel to Kyoto and summon the witness there because it is "between Yawata in Kyushu, where he [the witness] lives, and Tokyo [where the culprit himself lives]."[48]

Did Matsumoto know that the *imobō* of Hiranoya, especially the *ebi-imo*, had its origins in the Chinese taro of Kyushu? Were these two different encounters really coincidental after all?

Notes

1. Matsumoto, Seichō, "Kao" (The Face) (*Matsumoto Seichō zenshu 36: Chihō shi o kau onna* (The complete works of Seichō Matsumoto vol. 36: The woman who buys local newspapers), Bungeishunju, 1973, [first published in August 1956 in *Shōsetsu Shinchō*]), pp. 159–188. This quote is from page 184. The contents in square brackets have been added based on the *Matsumoto Seichō kessaku sōshu I* (Seichō Matsumoto masterpieces collection I) (Shinchōsha, 1993, p. 226). Other minor corrections have also been made; changes to the text are indicated in square brackets.
2. Motoyama, Tekishū, *Inshoku jiten jyōkan: a-so* (Encyclopedia of food and drink vol. 1: A to So), Heibonsha Library, 2012. There is no entry for *imobō* in Shiyō Makimura's *Osaka kotoba jiten* (The dictionary of the Osaka dialect), Kodansha Gakujutsu Bunko, 2012, nor in the work that it is based on Shiyō Makimura (Osaka dialect society), ed., *Osaka hōgen jiten* (The dictionary of the Osaka regional dialect) (Sugimoto Shoten, 1955).
3. The Japan Society for the History of Customs, ed., *Zusetsu: Edojidai shoku seikatsu jiten (shinsō ban)* (The illustrated encyclopedia of dietary life in the Edo period (New Edition)), Yūzankaku, 1996, p. 260.
4. Akiyama, Tomiko, Ōmura, Shige, Hirayama, Chizuru, *Obanzai, aki to fuyu: Kyō no daidokoro saijiki* (Obanzai, Autumn and Winter: A literary calendar of a Kyoto kitchen), Kawade Shobō Shinsha, 2020, p. 10.
5. Akiyama et al., *Obanzai, aki to fuyu*, pp. 23–24.
6. Takagi, Shirō, *Kyō no aji* (Flavors of Kyoto), Rokugatsusha, 1956, p. 56.
7. Usui, Kinosuke, *Kyoto mikaku sanpo* (A gustatory stroll through Kyoto), Shirakawa Shoin, 1962, p. 24.
8. Harada, Yosamatsu, *Bai bai hitori annai* (A one-man guide to buying and selling), Harada, Yosamatsu, 1878, p. 14. Ishida Kyokuzan ed. *Kyoto meisho annai zue* (Illustrated guide to famous places in Kyoto), Shōhōdō, 1887, p. 46.
9. Ishida, *Kyoto meisho annai zue*, p. 48. In addition, in Tsujimoto, Jisaburō, ed., *Kyoto annai miyako hyaku shu (zōho ni ban)* (Guide to the one hundred faces of Kyoto (Enlarged second edition)) (Shōtokukan, 1894, p. 208), "Kaiseki cuisine" includes Hiranoya, lower Gion Maruyama."
10. There will be numerous other references to *sekigashi* herein. For details, see Kato, Masahiro, ed., *Modan Kyoto: "Yūraku" no kūkan bunka shi* (Modern Kyoto: A spatial cultural history of "Yuraku") (Nakanishiya Shuppan, 2007).
11. Kato, Sadayoshi, *Kyoto meisho annai* (Guide to famous places of Kyoto), Murakami Kanbei, 1893.
12. *Rekishi shashin* (Historical photographs) (No. 160, 1926, p. 6), presents a photo of Suizan Kurokawa's photo "Chrysanthemums of Gion Hiranoya." The ceiling and walls are covered with reed screens to surround the flower bed, and stools are placed on the floor. This photo can also be viewed at the Kyoto Prefectural Museum of Kyoto Studies and Rekisaikan Digital Archive.

13. Kato, Masahiro, "Rakutō no yūenchi: Shukusai kūkan o enshutu suru butai" (Amusement parks in eastern Kyoto: Stages creating festive spaces) (Uesugi Kazuo and Kato Masahiro eds. *Chizu de tanoshimu Kyoto no kindai* (Enjoying modern Kyoto with maps), Fūyusha, 2019), pp. 134–139.
14. Kunikida, Doppo, "Saikyō ryōri shirōto hyō" (Amateur reviews of the cuisine in western Kyoto) (*Kunikida Doppo zenshū: dai 9 kan* (The complete works of Doppo Kunikida, vol. 9), Gakkensha, 1969), pp. 690–695. The quote is from pages 694–695.
15. Honda, Tokujirō, "Kyoto ryōri no tokushoku" (Characteristics of Kyoto cuisine) (*Fujin Sekai: Rinji zōkan* (Fujin Sekai: Extra edition, vol. 3, no. 6, 1908), pp. 33–44. Quotes are from pages 33 and 34. The Tokyo restaurant opened in 1872 and was in Hibiya (1–3 Yūrakuchō, Kōjimachi-ku). Okuda Udon (Okuda Kenjirō), *Shoku angya (Tokyo no maki)* (Food pilgrimage (Tokyo vol.)), Kyōbunkan, 1925, pp. 158-160.
16. Tanaka, Ryoko, "Warera no Kyoto 1" (Our Kyoto vol. 1) (*Kyoto*, no. 18 (vol. 3, no. 6), 1938).
17. Honda, "Kyoto ryōri no tokushoku," p. 34.
18. Ōkubo, Tsuneji, *Umaimon jyunrei* (Pilgrimage of delicious things), Rokugatsusha, 1958, p. 30.
19. Inoue, Jinnosuke, "Rakumi hanjōki – Kyō no tabemono, sono 3" (Rakumi prosperity chronicles: Food in Kyoto, Part 3) (*Rakumi* vol. 71, 1958), pp. 88–92. Quote is from page 89.
20. Yoshida, Hatsusaburō *Kyō meibutsu Imobō Hiranoya o chūshin to seru Higashiyama kara Eizan e no kōraku goannnai* (Guide to excursions from Higashiyama to Eizan, focusing on Hiranoya's famous Kyoto dish of imobō) Hiranoya Imobō Honten, August 20, 1928.
21. Takagi, *Kyō no aji*, p. 56.
22. Ōkubo, *Umaimon jyunrei*, p. 28.
23. Ōkubo, p. 28.
24. However, after Takagi and Ōkubo, Shishi Bunroku said in 1968, "In recent years, the seasoning has become very sweet, and footfall has decreased." *Shokumi Saijiki*, Chūko Bunko, 2016, p. 50.
25. Takagi, *Kyō no aji*, p. 55.
26. Inoue, "Rakumi hanjōki, sono 3," p. 88.
27. However, the address of the former is given as, 362 Gion, north side. Higashiyama Cuisine and Restaurant Association, *Sōritsu 100 shūnen kinen shi* (100th Anniversary Magazine), Higashiyama Cuisine and Restaurant Association, 2018, p. 105.
28. Kosuge, Keitarō and Yoshino, Hisakazu, *Kyoto shō kō jinmei roku* (Records of famous Kyoto merchants), Gōshi Shōhō, 1905, p. 94.
29. *Kyoto shō kō jinmei roku*, Kyoto Chamber of Commerce, 1907, p. 106.
30. *Yomiuri Shimbun*, April 10, 1906.
31. Dōmoto, Kansei, "Meiji no koro no Kyoto ryōri ya" (Kyoto restaurants in the Meiji era) (*Rakumi* vol. 100, 1960), pp. 34–36. Quote from page 36.
32. Yamanaka, Shōtarō, ed., *20 seiki no Kyoto: Ten no maki* (Kyoto in the 20th century: vol. 1), Kyoto Shuppan Kyōkai, 1908, p. 54.
33. Kosuge, Keitarō and Fujii, Yoshihisa, eds., *Kyoto shō kō jinmei roku; Taishō 4 nen kaisei* (Records of famous Kyoto merchants 1915 revised version), Gōshi Shōhō Kaisha, 1915, p. 26.

Teru Kitamura, who ran the Hiranoya branch, is also mentioned in the 1918 revised edition under "Yaoc, Upper Shijō, Pontochō." Thus, he probably also ran a teahouse in Pontochō. Kosuge Keitarō and Fujii Yoshihisa, eds., *Kyoto shō kō jinmei roku; Taishō 7 nen kaisei* (Records of famous Kyoto merchants 1918 revised version), Gōshi Shōhō Kaisha, 1918, p. 29.

34. Takeuchi, Yoshihisa, Kosuge, Keitaōo, Uda, Shūji, eds., *Kyoto shō kō jinmei roku; Taishō 13 nen kaiban* (Records of famous Kyoto merchants 1924 revised version), Kyoto Shōkōjinmeiroku Shuppansho, p. 27.
35. Kyoto Federation of Food and Beverage Industry Associations, *Kyoto ryō in 70 nen enkaku shi* (70-year history of Kyoto food and drink), Kyoto Federation of Food and Beverage Industry Associations, 1993, pp. 58–59.
36. On June 1, 1928, the owner and others were killed and injured when a resentful former employee set the restaurant on fire and burned it to the ground. *Kyoto Nichinichi Shimbun,* June 2, 1928.
37. Inoue, "Rakumi hanjōki, sono 3," p. 88.
38. Kokubu, Ayako and Nishiyama, Jirō (photograph) *Kyō no okami tachi: Shinise no aji o tazunete* (Kyoto hostesses: Visiting the long-established restaurant), Shibata Shoten, 1980, pp. 160–164. Quote is from page 163.
39. The pamphlet owned by the author says "Hiranoya Saryo." Kokubu and Nishiyama, *Kyō no okami tachi*, pp. 168–169. Evening Kyoto Shimbun, Gakugeibu ed., *Kyō aji 100 sen* (100 selections of Kyoto flavors), Tanko Shinsha, 1959, p. 161.
40. Yonehara, Yūji and Fujita, Akari, *Kyoto shinise: Noren no kokoro* (Kyoto's Old Establishments: Heart of the Noren), Suiyōsha, 2008, pp. 42–43.
41. Matsuzaki, Tenmin, 'Keihan tabe aru ki' (Osaka and Kyoto food diary) (*Santo tabe aru ki* (Food diary of three cities), Seibundō, 1932), pp. 1–128. Quote from page 21.
42. Takagi, *Kyō no aji*, pp. 55–56.
43. Dōmoto, "Meiji no koro no," p. 36.
44. Ōkubo, *Umaimon jyunrei*, p. 29.
45. Inoue, "Rakumi hanjōki, sono 3," p. 89.
46. Evening Kyoto Shimbun, Gakugeibu, *Kyō aji 100 sen*, p. 160.
47. Evening Kyoto Shimbun, Gakugeibu, p. 161.
48. Matsumoto, "Kao," p. 173.

10.2478/9784991351600-009

Chapter 2
Saltwater and Freshwater Fish Dishes in an Inland City

> Although far from the sea, the cuisine of Kyoto has come to include dried fish from Hokkaido. Buckwheat noodles with herring and *imobō* are examples of this, as indeed are herring and cod... In the same way that the vegetarian cuisine [*shojin ryōri*] of China has become completely assimilated here, the people of Kyoto seem to have a liking for ingredients from remote areas. Or it may be because of the pride of the Kyoto people, who want to show their unique characteristics separate from the ingredients used in neighboring lands. (Seichō Matsumoto, *Kyoto Cuisine*)[1]

2.1 Flavors unsuited to the palates of the people of Edo/Tokyo

2.1.1 Niken Chaya on the Kamiya River

In Kyoto, the old capital and a sightseeing city during the Edo period, teahouses offering food were almost always attached to famous sights to entertain visitors. One example of such an establishment is Niken Chaya, situated within the precincts of Hirano Shrine on the left bank of the Kamiya River.[2]

In 1789, on a recreational trip to Kyoto, Shiba Kōkan described the scene from Kitano Tenmangū Shrine to Hirano Shrine: "Exit the north gate of Kitano Tenjin Shrine, and one will find Niken Chaya in the valley by the river. They serve carp soup and grilled eel. From here are the three shrines of Hirano."[3] Shiba Kōkan was a keen drinker, so doubtless he would have enjoyed a bowl of grilled eel and sipped his "carp soup" with a cup of sake to hand.

The two main characters of the novel *Tōkaidōchū Hizakurige* (also known as *Shank's Mare*) (1804), Yajirobei (Yaji) and Kitahachi (Kita) also rest at Niken Chaya on the Kamiya River:

> Here was Niken Chaya, on the banks of the Kamiya River. They were both hungry, so they went to the teahouse to eat, whereupon a woman came out and greeted them, "How nice of you to come. Please come right inside."

> Yaji said, "Perhaps you have something delicious. We wish to eat food and drink sake. Please give us something small and quick."
>
> They seated themselves at the far end of the veranda, and the woman brought a sake vessel and cups. The side dish was dried *ayu* [sweetfish] simmered in soy sauce.
>
> Yaji said, "Thanks for coming so quickly. Please serve sake for us."[4]

Yaji and Kita are delighted, eating their dried sweetfish simmered in soy sauce, as the waitress serves sake for them. The ingredients offered at Niken Chaya include carp, eel, and sweetfish. Freshwater fish are variously made into clear soups, grilled, and simmered in soy sauce. The place offers typical dishes for a bar in an inland city. This chapter will provide an overview of the surprising diversity of "sea and river fish dishes" in modern Kyoto, paying attention to the locations of the restaurants.

2.1.2 Food reviews by Edo natives

> Three good things in Kyoto: girls, the water of the Kamo River, temples and shrines. Three bad things: the meanness of the people, the food, and the boat service.
>
> Five things that are lacking: fish, beggars, good medicinal infusions, good tobacco, and real prostitutes. (1802)[5]

> Having stayed in Kyoto for a long time, what one finds most inhibiting is the lack of Tokyo-style women and food. This is made all the more regrettable by the fact that the sake is so good. (1912)[6]

Here, we have two passages comparing Edo/Tokyo with Kyoto, in the context of arguments comparing the cities or cultures. The first is from a travelogue from 1802, which states, "I was born in Edo 36 years ago, and for the first time this year, I went to Kyoto, and I was exposed to the vulgarity of the people for a time." The second passage is from a diary written in 1912 by an "Edo native," who states that he was "setting foot in the lands of the Western Capital [Kyoto] for the first time in his life." Both are written by well-known writers, the first from Bakin Kyokutei (Takizawa)'s *Kiryo Manryoku* ("Miscellaneous Observations of a Traveler"), and the other from Junichirō Tanizaki's "Suzaku Diary." It seems that those from elsewhere who visit Kyoto want to compare not only the scenery and food but also the people and customs.

Both *Kiryo Manryoku* and the "Suzaku Diary" are more than mere records of time spent sightseeing in Kyoto as both are informed by a keen eye for observation. However, the two authors express quite different opinions of the same phenomena, perhaps because there is more than a hundred years and the epoch-changing Meiji Restoration between them, or perhaps because the tastes of the two writers differ.

A good example of this is their assessments of women (or "girls"). In contrast with Bakin Kyokutei listing "girls" as one of the three representative "good things," Tanizaki unhesitatingly laments the lack of "Tokyo-style women," saying, "when

I happen to travel to a strange land, what I miss the most is women and food."[7] Moreover, although Bakin criticizes "meanness of the people" of Kyoto (i.e., their stinginess), Tanizaki defends them elsewhere in his diaries, saying: "the character of the Kyoto people is by no means a bad one."

However, they agree that the food is unpalatable, with Bakin referring to "the food" as one of the "bad things" and "fish" as one of the "things that are lacking." Let us examine his critique a little more closely:

> The *ikesu* [see below] are mainly close to the Takase River, where the air is fresher in the summer. Kashiwaya and Matsugen are popular. Kashiwaya also has a branch in Pontochō. Matsugen has had many customers in recent years. Its specialties are said to be eel, conger eel, and carp. The fish served are salted sea bream and salted abalone from Wakasa, carp and crucian carp from Ōmi, and other fish from Osaka; many of these spoil in the summer. Many of the eels and congers come from Wakasa. However, they are very oily and inferior to those of Edo. The sweetfish and minnows caught in the Kamo River are thin and bony. The minnows are good. One may say that the grilled sweetfish of Wakasa are good, and yet seem tasteless in the mouth of a person accustomed to eating river fish of the same age from Gifu. The *koikokushō* [carp in miso soup] is also made with white miso. There is no red miso. The white miso is too sweet and not salty. *Dengaku* [miso glazed] dishes are also served with this white miso, which is difficult for Edo people to eat... The only tasteful things in Kyoto are gluten, *yuba* [tofu skin], taro, mizuna, and udon noodles, and the rest are *not suited to the tastes of Edo people.*[8]

The word "ikesu" refers to restaurants that mainly served freshwater fish and were mostly located facing the Takase River between Sanjō and Shijō roads. According to Bakin, long fish such as eels and conger eels were typical items from the rivers and seas, and *koi-no-arai* [carp sashimi] seems to have been the local specialty.

In Kyoto, saltwater and freshwater produce was brought in from distant Wakasa Bay and Osaka Bay, and nearby Lake Biwa. Bakin finds the sea fish to be inferior to those of the Edo region and finds the local specialty of sweetfish to not be in the same league as those from the Nagara River.

Then, as now, the yuba and udon are described as delicious,[9] but Bakin concludes with a dry comment that nothing else is suited to "the tastes of Edo people."

Tanizaki, like Bakin, is similarly unforgiving:

> The food of Kyoto is bland and watery, not likely to suit the tastes of Tokyo people. First of all, the taste of the soy sauce is very different. Although each person may have their own tastes, the eel, sushi, soba, and so on are far inferior to those of Tokyo. There is generally little variety of sea fish, which moreover seem to be poor in quality.[10]

Tanizaki the Tokyo man, like Bakin, finds not only the white miso to be not to his liking but also the soy sauce, sushi, soba, and fish. Over a hundred years had passed since Bakin's time, and yet the views of Edo/Tokyo people appear to be unchanged.

However, Tanizaki also remarked that "people have their likes and dislikes, but first of all, the very best in Kyoto should not be disgraced" and, specifying the "cuisine of Hyōtei," recommended that "everyone, even Tokyo people, should eat this cuisine."[11] Below is a description of Tanizaki's visit to Hyōtei with poet and playwright Mikihiko Nagata at the invitation of the writer Bin Ueda:

> For a time, the rain became stronger again, enveloping the surroundings of the tastefully designed teahouse in 10 or 20 layers of downpour, beating the pond water, washing the green moss, and blanketing the modest garden in a thick layer of mist. Enjoying the sound of the water droplets running around the pipe, creating a pleasant sensation as though they were sinking into my intestines, I took a sip of the warm yet not too warm sake, feeling it against my tongue.
>
> I was neither hungry nor too full, in perfect spirits to be invited to such a place. First, a small plate of tofu is served, cut into pieces about the size of the *ankake* [tofu topped with bean paste] of Sasa-no-yuki. The tofu is topped with a thick sauce, green and white. Although it is not as soft as the silken tofu of Tokyo and has a different texture, when you put it in your mouth, it has a completely different and unforgettable taste. The sauce is a refined, sweet, and beguiling liquid, perhaps made of powdered *ki-no-me* [young Japanese peppercorns] dissolved in sugar. Starting with minnows from the Kamo River that have been cooked in oil, countless other freshwater fish and vegetable dishes with unknown names that seemed to be specialty produce of Kyoto, were served one after another, using the signature tableware of the establishment. From the side, the maids came to clear away the empty plates and lidded bowls, changing the chopsticks one by one and withdrew without pouring any sake. The sake was good, so I was able to eat as much as I desired without the alcohol getting to my head.[12]

In the latter half of Tanizaki's stay, the writing style of the *Suzaku Diary* becomes somewhat disordered, but the scenes at Hyōtei are described with crisp clarity. On this occasion, Tanizaki was eating a river fish that seemed to be a "specialty produce" of Kyoto – "minnows" (*higai*) from the Kamo River. We wonder what brand of sake he was drinking.

In Tanizaki's novel *The Makioka Sisters*, the family of the protagonist has a long-standing custom of viewing the cherry blossoms in Kyoto and lunching at Hyōtei.[13] The quotation above describes Tanizaki's first experience at Hyōtei.

2.2 Saltwater fish cuisine and the "True flavor of Kyoto"

2.2.1 Mackerel sushi and conger eel

"I always say that *imobō*, mackerel sushi, and herring soba are representative of the true taste of Kyoto, which cannot be found anywhere else," writes Kinosuke Usui in *A Gustatory Stroll through Kyoto*, which greatly influenced the writing of this book.[14]

You may notice that the three dishes identified as the true taste of Kyoto – "*imobō*, mackerel sushi, and herring soba" – have one thing in common: saltwater fish.

As noted in Chapter 1, the "bō" part of "imobō" is derived from "bōdara," which refers to dried cod produced in Hokkaido. If we now skip to the last item in the list above, the herring used in "herring soba" (*nishin soba*) is also a fish mainly offloaded in Hokkaido. According to Matsuba, an establishment next to the Minamiza Theater, it developed herring soba made using dried and sliced herring in 1882.[15] Eighty years after it first became available, it came to be described as one of the true flavors of Kyoto. Today, it is so widespread that it features on the menus of udon restaurants in the city. *Nishin* [herring] *soba* may rightly be described as a great invention.

In the early twentieth century, there was a specialty herring restaurant in Kyoto that offered "simmered herring, grated yam soup, *umewan*, a cup of sake, pickles, and rice, all for 90 *sen*, and that was known as a herring restaurant that couldn't offer anything else."[16] This was Kyōsenryō, located in Makatsugahara, which has been famous since the Edo period. It was said that "herring has long been part of the daily diet of the people of Kyoto, and something with which they are deeply familiar, to such an extent that it is said that a woman's adeptness at cooking herring was one of the things that determined her suitability as a wife." In this context, it must take some confidence to run a restaurant serving only herring dishes.[17] Indeed, despite much research to find other references to restaurants of a similar kind, there seemed to be no other such restaurant in Kyoto; it appears to have been the first and last of its kind in history.[18]

The final item in the list above, mackerel sushi (*sabazushi*), is made from mackerel that was caught and salted in Wakasa before being transported to Kyoto via the so-called "Mackerel Road" (Saba Kaidō). Kinosuke Usui explains:

> When talking about the sushi of Kyoto, one should not overlook the mackerel sushi of "Izū."
>
> In the olden days, mackerel caught in Wakasa was immediately salted and brought to Kyoto; in just a day and a night, the salt would have worked, and it would be ready to eat. This would then be made into *saba-no-bōzushi* [rod-shaped mackerel sushi] through skillful use of the salted mackerel.[19]

In this way, the true flavor of Kyoto is seen as resulting from the clever processing and preservation of ingredients created from fast-moving sea fish. Mackerel sushi is part of the traditional foodscapes associated with religious festivals, but the traditions of Osaka and Kyoto differ significantly in this regard, as we can see in Shimada Fukuo's description:

> Even the treats for the same summer festival are very different in Kyoto and Osaka.
>
> At the Tenjin Festival in Osaka, you would have teriyaki conger eel with chilled *sōmen* noodles, while at the Gion Festival in Kyoto, you would have things like mackerel sushi.

> Kyoto's mackerel sushi is much more dignified than Osaka's, which they call *battera* or *matsumae sushi*. Secondly, it is sliced very thick. Thirdly, the *kobu* [kelp] wrapped around it is white in Osaka, but black in Kyoto ... so it's different.[20]

Mackerel sushi is the definitive festival food in Kyoto. Even the city's udon restaurants receive orders for mackerel sushi on festival days. However, this is not to say that conger eel (*hamo*) is not eaten in Kyoto. In fact, conger eel is served everywhere in Kyoto in the summer, from high-end Japanese eateries to bars. According to Jinnosuke Inoue:

> If there were a most popular dish on Kyoto tables in summer, it would undoubtedly be the ubiquitous conger eel. Whether one finds it delicious or unpleasant, one may say that the people of Kyoto become accustomed to eating the conger eel from childhood.[21]

Conger eel feed and grow large during the rainy season. Although a high degree of skill is required to remove the bones, they form a key part of summer in Kyoto and can be found in various forms, from the expensive *yubiki hamo* (parboiled conger eel) and *yakishimo* (skin seared) at traditional Japanese restaurants to the somewhat smaller pieces of tempura found at *izakayas* (traditional Japanese bars). It is another indispensable saltwater fish.

Furthermore, Kinosuke Usui notes that conger eel has certain advantages over mackerel sushi:

> Also, mackerel is not in season around the time of the Gion Festival in summer, and it is a little hot for mackerel sushi, so it is better to switch to conger eel sushi. This substitutes the Kyoto specialty of grilled conger eel for salted mackerel; it is more elegant than mackerel sushi and well liked for the coolness it brings in summer.[22]

Thus, "hamo-zushi" (conger eel sushi) is another "Kyoto specialty."

2.2.2 Oyster boats on Lake Biwa Canal

The only current restaurant that comes to mind that serves a specific type of seafood is Hibanaya, a restaurant specializing in *fugu* (blowfish) cuisine in Takano in the Sakyo Ward of Kyoto. Hibanaya has a long history with sources indicating that it was founded in 1919.[23]

In the modern history of Kyoto, as well as the herring specialist restaurant Kyōsenryō, oyster restaurants have occasionally emerged, but they remain unusual. In October 1886, "seeing that there were still no restaurants serving raw oysters, a specialty of Hiroshima Prefecture" in Kyoto at the time, the oyster specialty restaurant

Yaomasa was founded in the Saiseki area of Kyoto. Yaomasa later moved to the west end of Shijo Ōhashi Bridge and is now called Tōka Saikan.

Doppo Kunikida, a novelist and journalist of the Meiji period, refers to the "oyster dishes of Yaomasa" as follows: "Upon approaching the Shijō Bridge (western end), there is a three-floor restaurant called Yaomasa, which serves over 10 kinds of oyster dishes, including oyster rice, oyster tempura in winter, and oyster tofu *karaae* [mixed with soy pulp]."[24] In 1907, a northern branch (Kitaten) was opened on the north side of Shijō-dōri road,[25] mainly focusing on oyster dishes (Fig. 2.1), whereas the main restaurant (Honten) gradually specialized in Western-style cuisine. Unfortunately, Yaomasa did not survive the war. Another notable oyster restaurant in the early Shōwa period, Kakiharu, claimed to be the "original oyster boat" restaurant. The following quotation is a description of the location of Kakiharu from Naoya Shiga's novel *A Dark Night's Passing (Part Two)* published in the 1920s in the magazine *Kaizō*:

> After crossing the bridge, the two of them took a train from there to Shijō and then went to an oyster boat at the side of a narrow bridge called Kikusuibashi. It was Kensaku's first time on an oyster boat since Onomichi. Hence, the painful memories of that time flitted through his mind, but those memories had since become veiled, and he was currently happy. For one, the atmosphere of the place he was in now was completely different. This place bore little similarity to the oyster boat in that dingy warehouse district. Before him were the lights of the many teahouses of Gion. The lights of the gaudy Shijō Bridge, and beyond it the Minami-za Theater, shone and reflected with a dazzling insistence against the waters of the river.[26]

Figure 2.1 Nighttime view of Yaomasa (Kitaten and Honten)

"Oyster boats" (*kakibune*) were a unique form of restaurant, originally seen as a specialty of Osaka.[27] During the oyster season in late fall, oysters would be transported from Hiroshima by boats which would moor in the waterways of Osaka, such as Dōtonbori and Tosabori, and food would be served on board. From the 1910s, the oyster boats started operating as restaurants throughout the year, serving river fish dishes as well as the oysters that were available in winter. Kakihiro, which is still located by Yodoyabashi Bridge on the banks of the Tosabori River in Osaka, is a remnant of that time.

Unlike central Osaka, there are no rivers in Kyoto that lead directly to the sea, and no oyster boats come upstream. In *A Dark Night's Passing*, the newlyweds Kensaku and Naoko, are able to see the lights of the teahouses in the Gion Kagai district, as well as the Shijō Ōhashi Bridge and the Minamiza Theater from the oyster boat. Where were they located to have such a view...?

Screenwriter Yoshikata Yoda (1909–1991), known for his work with director Kenji Mizoguchi (1898–1956), provides a clear explanation of the location of Kakiharu:

> Cross the Shijō Ōhashi Bridge, heading east. From the Keihan stop, walk up the riverside. Against the backdrop of the rich waters of the canal flowing alongside the Kamo River, 10 or so willow trees stand alongside oden and yakitori stalls with red lanterns hanging from them. After this is Kakiharu, a boat-shaped restaurant that overhangs the canal.[28]

Thus, the "oyster boat" Kakiharu was not literally a boat that floats on the water but a restaurant in the shape of a boat overhanging the Lake Biwa Canal, and proud of its "home-reared oysters and fresh seafood cuisine."[29] Yoda praised the restaurant saying, "it gladdens me that I can eat fresh, clean *kaki-nabe* [oyster hotpot], which is made by pouring clear *dashi* stock broth into the pot, so that one can see the green of the Japanese parsley and the white skin of the grilled tofu." However, he did complain that "the oyster rice that is served seems somewhat heavy after eating a large meal."

An oyster boat that floats on the canal, or rather, appears to float on the canal seems highly characteristic of an inland city. Kakiharu also had a branch on the Misosogi River near the Kaburenjō kabuki theater in Pontochō,[30] meaning that the restaurants were strangely positioned on both sides of the Kamo River. Speaking of land-based oyster boats, we may be reminded of the restaurant Kakibune (literally "oyster boat") located near Matsumoto Castle in Nagano Prefecture. Although there are oyster bars in modern Kyoto, there are probably no restaurants specializing in oyster cuisine.

However, the advertisements from *Rakumi* quoted above state things like "Oysters served from October 1st" and "The long-awaited season has finally begun," indicating that oyster dishes were strictly seasonal. For this reason, Kakiharu also served "saltwater and freshwater fish dishes" (Fig. 2.2).

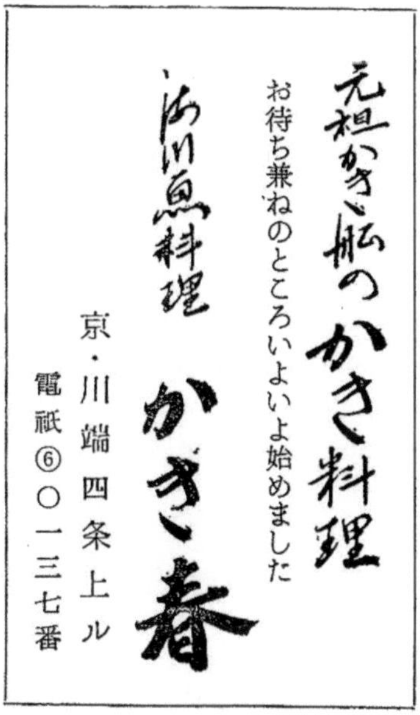

Figure 2.2 An advertisement for Kakiharu

2.3 A diversity of freshwater fish cuisine

> Every summer, I am reminded of the simple taste of *gori* cuisine served when I visited Kifune in days gone by.[31]

2.3.1 Rosanjin's praise

The guidebook *Daikyōto* (Great Kyoto) published in the mid-1920s contains a section on "specialties and souvenirs of Kyoto." Alongside more familiar Kyoto products including confectionary such as *yatsuhashi* and moon cakes and pickles such as *senmaizuke* and *suguki*, the last item listed was something called *sagishirazu* ["unknown to herons"], a word that the authors had never seen before, although it was apparently well-known. Below is a quote from the introduction to *Daikyōto*:

> Small fish called "gori" are a product of the Kamo River. They are simmered with soy sauce and sugar to soften the fish bones, eliminate any fishy odor, and make them able to withstand long-term storage. *They are prized as one of Kyoto's specialties.* The name "sagishirazu" comes from the fact that even herons and other creatures that catch and eat small fish will overlook fish as small as these.[32]

So, *sagishirazu* refers to a fish called "gori" that is so small that even herons will not prey upon it. But in Kyoto, there is a clear distinction drawn between *sagishirazu* and *gori*. The artist Rosanjin Kitaōji observed that "Kyoto has a delicious fish called '*sagishirazu*,'" and he praised *gori* as well, distinguishing it from *sagishirazu* as follows:[33]

> In the freshwater fish cuisine of Kyoto, there is a custom of putting seven *gori* in a bowl of *akadashi* (miso soup). You only need to add seven of such a small thing in order to make such a *wonderful Kyoto specialty soup*, so you can imagine how delicious those little fish must be. This also makes them pricey because they don't catch many of them. They don't even catch enough to do things like preserving them in soy sauce. But people still want to eat them preserved in soy sauce, so *gori chazuke* [tea rice soup of *gori*] is said to be the most luxurious thing in all creation.[34]

In this way, Rosanjin also recognized this as a "Kyoto specialty."

According to an acquaintance who works at a long-established freshwater fish shop in Nishiki Market, known as "Kyoto's kitchen," *gori* refers to small *yoshinobori* (freshwater goby) and *sagishirazu* refers to *oikawa* (freshwater sprat) fry. The store sells *gori* preserved in soy sauce, but they use a small fish called *urori* from Lake Biwa. The *gori* caught in the river is sold under the name "hongori" ("true gori") to distinguish it from *urori*, and it appears to be popular when sold alive wholesale to restaurants.

Kansei Dōmoto recalls that "there were several restaurants specializing in freshwater fish" in Pontochō and Nawate-dōri in "an era when delicacies from the Kamo River such as *gori* and *sagishirazu* were prized."[35] In fact, the *Illustrated Guide to Famous Places of Kyoto*, published in 1887, includes a section on "Sagishirazu Merchants," which lists four such establishments, including some on Nawate-dōri.[36] The upper right corner of the illustration for this section has the words "catching *sagishirazu*, Kamo River" (Fig. 2.3). The left bank of the river is depicted in the background, and the fisher seems to be casting their net in shallow water with a gentle current. It was such a popular product that specialty stores were established.

Such was its popularity that *gori* featured on the menu of Heihachi Jaya, a long-established restaurant renowned for its freshwater fish dishes.[37]

> It retains the vestiges of a roadside teahouse of old, where travelers from Wakasa and the processions of feudal lords traveling to and from Edo under the system of *sankin-kōtai* [alternate attendance] would stop to rest. This establishment would make grated taro harvested from nearby mountains and serve it to such travelers with grilled or simmered freshwater fish and barley rice. It gladdens me that this nostalgic flavor has been handed down to the present and made into specialty dishes. Its rooms have a view of the Takano River across the garden, and the food is brought to you by servers wearing sashed *kongasuri* [dark-blue cloth with white splash patterns] garments and *mihabamae-dare* [broadcloth aprons]. Freshwater fish dishes such as *higai*, *moroko*, river prawn, *gori*, *amago* [red-spotted masu trout], eel, and carp; the restaurant's most prized dish is *gori amedaki* [*gori* stewed with soy sauce and sugar] made personally by the proprietor.[38]

Figure 2.3 A scene of fishing for *sagishirazu*

The writer Sōseki Natsume, visiting Kyoto in the spring of 1907, wrote in his diary (dated April 10): "Heihachi Jaya (Braved the rain; rented a car with Kyoshi. Streams, mountains, carp broth, and eels)."[39] The entry from the previous day also reads: "Jyūichiya. Heihachi Jaya. At the edge of a mountain on the way to Kōya Village." Sōseki visited Heihachi Jaya at the invitation of the poet Kyoshi Takahama, who had just arrived in Kyoto.

We can envision the two enjoying freshwater fish dishes in a tatami room facing the clear waters of the Takano River (Fig. 2.4). Perhaps, Sōseki ate *gori* without knowing it.

Figure 2.4 A tatami room in Heihachi Jaya facing the Takano River

2.3.2 Catfish, loach, and soft-shelled turtle

Sōseki's diary entry from April 9, 1907 mentions "Jūichiya," which was a rare kind of restaurant specializing in "catfish cuisine." Like Heihachi Jaya, it seems that it was originally a roadside teahouse. It was "commonly referred to as 'Samurai-ya'" because warrior monks (temple priests who were members of the samurai class) would come down from Mount Hiei to rest there. After the samurai ceased to exist, the single kanji that had been used to represent their class (士) was broken down into its constituent elements, which are the characters for "ten" (十) and "one" (一), from which the name "Jūichiya" (literally meaning "ten-one restaurant" or "eleven restaurant") was derived.[40]

The menu included "a complete set of catfish dishes." Among these dishes, Tsuneji Ōkubo, who was known as a food connoisseur, praised the *namazu-no-arai* (catfish sashimi in chilled water) as "resembling carp, yet being far more delicious than the carp version of the dish, the catfish flesh being deep red in some places and white in others."[41] According to Ōkubo, in the late 1950s, the restaurant purchased catfish locally and "kept them in its own pond." However, when the author (Kato) interviewed the original owner of the restaurant in 2010, he said that in later years, the catfish were purchased from Okayama. The refrigerator that was still used in the kitchen used ice instead of electricity, which was perfect for keeping sashimi cold. Kunio Yanagita pointed out the connection between the spread of refrigerators and the increased opportunities – or the growing desire – to eat raw sea fish as sashimi; this also extends to catfish dishes.[42]

In addition to catfish, there were restaurants that specialized in loach. Although eel restaurants also served loach (Fig. 2.5), only one loach specialty restaurant has existed in Kyoto throughout much of the modern era: Minoshō, located in front of the Imperial Palace gates on Sakaimachi street.

Figure 2.5 Advertisement for Umenoi indicating that both eel and loach were available

In a guidebook published in 1881, the "eel cuisine" section features Minoshō (on the intersection between the Takakura and Marutamachi roads) and another restaurant called Wakaebi (eastern end of Gojōbashi Bridge).[43] These restaurants may be regarded as specializing in loach cuisine as they are listed separately from the section on "*ikesu* freshwater and saltwater fish dishes."

> Loach has a long history in the form of "yanagawa-nabe" in the Kansai region. The body of the loach is cut open and simmered with burdock in an earthenware pot, and beaten eggs are dropped into the "yanagawa-nabe" or "dojō-nabe" [loach hotpot]. The dish has a slight sweetness and spiciness and is eaten seasoned with Japanese pepper. It is a truly unsophisticated dish but has an aspect of homeliness, and *it is something that all Kansai people will be familiar with, and will have been eating since childhood.*[44]

Thus, we see that loach hotpot was something that all Kansai people, including residents of Kyoto, would have been eating since childhood. Rosanjin Kitaōji describes it as follows:

> "Loach hotpot" – something tasty, cheap, nourishing, familiar, easy to make at home, and utterly beyond reproach. Yet it is not aristocratic. It is welcomed no matter where you go and rightly so.[45]

We may wonder whether there are still families today who eat loach hotpot with their children.

Minoshō was located at the intersection of Marutamachi and Takakura streets until the early Meiji period (mid- to late 19th century), before being relocated to Sakaimachi, one street to the east. Its location in front of the Imperial Palace was unusual, as were its opening hours, "it is also a strange restaurant that opens at 6:00 am and closes at 6:00 pm."[46] These are points of interest, alongside the restaurant's origins, which remain obscure.

Minoshō's reopening after the war was delayed until the autumn of 1956, making its fans nervous. The painter Shujaku Umezaki (1896–1969) provided the following account of his visit to Minoshō in the spring of the following year:

> Firstly, a chance reunion with the proprietress strengthened my initial impressions. A Gion native, she had often sat with her predecessor, the venerable Kansetsu, at drinking parties (which often literally went on for nights and nights). Naturally, the memory of her late mentor made the conversation very lively.
>
> The flavor of the loach cuisine there perfectly summons up memories of generations past, and of course, the ingredients are the produce of Okayama, the best in Japan. Small loaches are made into hotpot, and the larger ones are grilled. The menu also includes loach tempura, loach *maki* rolls, and loach soup.[47]

Advertisements from the time also promoted eel[48], so the restaurant wasn't exclusively devoted to loach (see also Chapter 3). All the same, it seems doubtful that anywhere nowadays would offer a menu solely comprising catfish followed by loach.

After catfish and loach, we have soft-shelled turtle (*suppon*). According to Frank Hawley:

> After it had been decided that I would come to Kyoto, Hideo Nagata, my gastronomic acquaintance, needed only to see my face, before he would urge me repeatedly thus: "Mr. Hawley, when you go to Kyoto, first and foremost, you must eat soft-shelled turtle." However, the "first and foremost" place for soft-shelled turtle cuisine was "Daiichi," which was somewhere in Kitano, on the edge of the city, distant from our residence. I was unable to go there for some time after arriving in Kyoto as I had been exclusively making a food pilgrimage centered on the nearby area of Gion.[49]

In Japanese, there is a pun in the advice that "first and foremost, you must eat soft-shelled turtle" because the phrase "first and foremost" is a homonym for Daiichi, the name of the restaurant. Frank Hawley (1906–1961) was a British linguist appointed to the Daisan Kōtō Gakkō (Third High School), which was established during the Meiji period and was a forerunner to Kyoto University. On his first visit to the restaurant, Hawley had no clear idea of where it even was; he "put on a brave face and hailed a taxi" telling it where to go by saying "Soft-shelled turtle – Daiichi!" Thereafter, they "passed through the Nishijin area, full of weavers, leaving the larger roads to enter winding narrow streets, before stopping in front of a small dimly lit house on a grubby alley on the edge of a brothel district!" He confesses that he was "more than a little taken aback." Upon revisiting the establishment, he said, "In general, houses in Kyoto are terribly gloomy as they have no gates or gardens, are fitted with latticework, have very narrow frontages, are strangely long in depth, and are jammed in closely against each other." His impressions, albeit in stark contrast to the "*Irikin*" style experienced by Tanizaki (discussed later), are generally accurate.

After this initial experience, Hawley seems to have visited many times over a short period, stating that "my greatest pleasure in Kyoto is not touring shrines and temples but walking around seeking delicious delicacies, so tasting these soft-shelled turtle dishes is a joy." He goes so far as to say that "this is the first place one is to take foreigners one is introducing [to Japan] from overseas and what pleases them above all else is the soft-shelled tortoise cuisine that can be found beyond this small, dirty entrance and nowhere else." He must have been very keen on it indeed.

According to the restaurant's owner at the time, "We only use soft-shelled turtle produced domestically. Those from Lake Biwa are first-class, and we mainly use them because of their geographical convenience."[50] But why was it located near the Gobanchō red-light district?

2.3.3 The city of sweetfish

For Rosanjin, who stated that "although the *ayu* [sweetfish] of Gifu is also famous, I can't say that it is the best *ayu* to my tastes," the sweetfish that he evaluated as being "number one" was that caught in the Wachi River, part of the Yūra River system in the northern part of Kyoto Prefecture: "It has a good shape, firm flesh, and a good aroma."[51] Other than the usual kanji for sweetfish (鮎), it can also be written as 年魚 (annual fish) or 香魚 (fragrant fish). Looking back on his childhood, Rosanjin explains the sweetfish situation in Kyoto as follows:

> As Kyoto is geographically blessed, since times of old it has been a custom for restaurants to serve *ayu* that are kept in individual family houses. Not only restaurants but also fishmongers sold live ayu to the general public.
>
> In our childhood, *ayu* would often be brought from the Katsura River in Saga to be sold in buckets with the fish dancing and flapping around. The fish has to be breathing to make the water dance like this, and if it couldn't breathe properly, the *ayu* would die in an instant. This became a special technique for selling *ayu*.[52]

Thus, it seems that there were traveling sellers of sweetfish employing sophisticated techniques. Rojinsan's account is from the Meiji era,[53] and of course this can no longer be seen in modern Kyoto. Today, according to the acquaintance at the freshwater fishmongers in Nishiki Market, summer is the main time that they "deliver live sweetfish." Sweetfish of different sizes are sorted according to customer loyalty, placed in a bag with water and oxygen, and delivered. Although the "special techniques" employed by the traveling sellers were not handed down unaltered, the "custom" of the restaurants, to use Rosanjin's term, are still alive today in Kyoto.

In his childhood, Jinnosuke Inoue said he had the "special technique" needed to bring live sweetfish from the Katsura River in buckets and that "when the season comes, all the restaurants are sure to serve salt-grilled *ayu*. It's safe to say that there wasn't a restaurant that didn't serve *ayu* dishes."[54] According to the acquaintance mentioned earlier, there are many restaurants that insist on having "wild sweetfish" and that some of the regular customers make a special effort to buy "live sweetfish" because they will be grilling it at home. Although there are no longer traveling sellers, specialty establishments that handle live sweetfish remain part of the local food culture.

Perhaps the most famous purveyors of sweetfish cuisine in Kyoto are Tsutaya and Hiranoya (a different restaurant than the previously discussed *imobō* Hiranoya), an inn and a teahouse, respectively, that serve sweetfish and can still be found today in Toriimoto near Mount Atago.

> Sweetfish specialty restaurants in Kiyotaki, such as Tsutaya and Hiranoya, may at first glance appear to have a whiff of the rustic about them, but they bring unaffected enjoyment to lovers of sweetfish. This area is commonly called "Toriimoto" as it is located close to Kiyotaki at the foot of Mount Atago, at the site of the first torii of Atago Shrine. Tsutaya is in front of the torii, and across from the torii is Hiranoya. Tsutaya is called the "sweetfish inn," and Hiranoya is called the "sweetfish teahouse." It is amusing to see the sight of these two establishments both competing with one another across the torii to invite people of the city in the midst of the quietness of the forest.[55]

The writer Tsutomu Mizukami stated that "Hiranoya is a sweetfish inn,"[56] but he may have misremembered and mistaken it for Tsutaya. Fellow writer Shōtarō Ikenami, who lived in the same era as Mizukami, added further praise that "Hiranoya's sweetfish is so delicious that it is just as though one were eating it in the [Edo atmosphere], as though one has wound back the days and months to a time a hundred years ago, so that one's sense of taste and traveler's spirit are harmoniously blended together into an indescribable feeling."[57]

> Hiranoya is an old teahouse dating back to the Kyōhō period [1716–1736], and people making a pilgrimage to Mount Atago would have rested there before ascending the mountain path.
>
> It is said that in the past, sweetfish caught in the Hozu and Kiyotaki rivers were brought to this teahouse, where the water was changed in the cargo of sweetfish before transporting it to Kyoto. The teahouse interior still retains strong vestiges of the Edo period. When people come here to eat and drink in the evening, the dim lighting gives one a sense of being in a dream.
>
> Although it is now several years ago, I once came here alone around noon and ate *koi-no-arai* [carp sashimi] and sweetfish grilled with salt across its back to my heart's content, sitting on a scarlet rug overlooking the valley.[58]

This is why Kyoto has been called the "city of sweetfish."

2.4 The disappearance of a genre

Wakasa Bay and Osaka Bay are both about 50 kilometers from Kyoto's central urban area. This basin, which borders the Lakeland of Ōmi, has great abundance of "freshwater and saltwater fish" brought from three directions. Around 1935, the foodscape of this area took on a new dimension. With the construction of national highways between Kobe, Osaka, and Kyoto, the time required to transport goods by road between Osaka Bay and Kyoto was greatly reduced, making it easier to distribute fresh fish.

As a result, Jūnidanya (which was located on the north side of the Gion district in this era), which is now known as a shabu-shabu and *ochazuke* (tea rice soup) restaurant, advertised that "we take absolute pride in our cuisine and specialize in cooking live fish from the Kyoto and Osaka area," and proclaiming that "we use live fish *delivered directly every morning from the very best producers in Akashi*."[59] Similarly, the menu at Tagoto, a venerable restaurant in Shinkyōgoku, stated that "*we receive fresh raw fish from Akashi every morning from Komagabayashi in the Kobe area*."[60] Komagabayashi is a fishing village near the Akashi Strait within the city of Kobe. Akashi sea bream and other fresh fish would have been transported to Kyoto. The reduction in the time and distance required for delivery stimulated demand for products with provenance and freshness. It was around this time that "Edo-style nigiri specialist" sushi restaurants began to appear in Kyoto.

2.4.1 The cosmology of the Kyoto Basin

Figure 2.6 is a sketch of the Kyoto Basin created by the authors based on the foodscape of the freshwater and saltwater fishes in that area. Although neighborhood children can sometimes be seen catching fish in the Kamiya River in the summer, the only large water systems with a significant fish presence are the Hozu River (Katsura River) and Kamo River. The village of Toriimoto, located at the foot of Mount Atago and at the first torii of Atago Shrine, is situated at the entrance to a valley. The teahouses here in Toriimoto played a key role in transporting and relaying sweetfish from the distant Yura River system to the urban area of Kyoto.

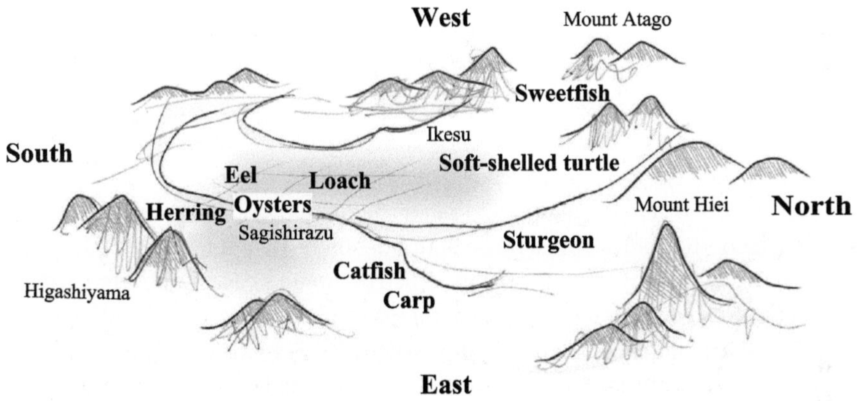

Figure 2.6 A cosmology of the Kyoto Basin

In Nishinokyō, on the western fringes of the old city, there were two restaurants that had been referred to as *ikesu* since the Edo period: Tsurikesu and Zaokesu. Unfortunately, this book is unable to touch on them further, but Tsurikesu, which is still in business today, has a very interesting history.

Located in Rokubanchō on the outskirts of the former red-light district Gobanchō, the soft-shelled turtle restaurant Daiichi still looks much the same today. However, how it came about that there is a restaurant serving such cuisine in this location in the first place remains a mystery.

The loach restaurant Minoshō's location in front of the Imperial Palace is also downright mysterious. There are neither any areas that produce tortoise or loach on the outskirts of the city, nor any conditions particularly suitable for raising them there.

The Kifune River is a tributary of the Kamo River; it flows by the approach to Kifune Shrine and is lined with ryokan with raised decks where one can enjoy a breeze in the summer. These establishments also serve freshwater fish to those seeking coolness in the summer.

> In the summer, the decking overhanging the stream is 5 or 10 degrees cooler than the city, making it a perfect spot for summering. In terms of food, young sweetfish and *haruko* [*amago*/trout] are eaten in the spring, and dishes to be enjoyed in the cool evening on the decking are served in the summer, including *amago* and sweetfish.[61]

Hopefully, it will be possible to take a closer look at the cuisine of the ryokan district in Kifune in another publication.

Turning our attention to the southwest of the city, Yase Amusement Park on the left bank of the Takano River at the foot of Shimeidake was described in the introduction. None of the restaurants that were once located there survived, except for Heihachi Jaya, which "specializes in freshwater fish including carp, crucian carp, eel, *amago*, *gori*, and *moroko* [gudgeon]."[62] The main restaurant of Heihachi Jaya is still located on the mountain's edge on the left bank of the Takano River, but Jūichiya, with its catfish cuisine, closed its doors long ago.

The area where the Takano and Kamo rivers meet is called the Kamo Delta. The area is strictly not a river delta at all, so its name probably comes from its shape, which resembles the Greek letter delta, rather than its topography. The famous entertainment districts of modern Kyoto are lined up on both sides of the river south of Kōjin Bridge. On the right bank, we have Higashi Sanbongi, which became a *sekigashi* table rental district after it became an entertainment district in the Edo period, the *sekigashi* district of Kamikiyamachi, the Pontochō entertainment district, the Saiseki restaurant district, the Shimogiyamachi *sekigashi* district, and, finally, the red-light district of Shichijō Shinchi. On the left bank, there is the red-light district Nijo Shinchi, and then the southern part of the freshwater fish restaurant district of Nawate-dōri, which was originally part of the Gion Shinchi area. From there, we continue to Miyagawachō, south of Shijō-dōri.

The restaurants specializing in *sagishirazu* on Nawate-dōri are Kyoto icons, and on the right bank, oyster and eel cuisines have been added to the lineage of *ikesu* cuisine dating back to the Edo period. Interestingly, the locations of restaurants that continue to preserve what Kinosuke Usui called "the true flavor of Kyoto" have not changed. These include the *imobō* of Hiranoya in Makuzugahara (Maruyama Park), *saba-no-sugata-zushi* (mackerel sushi) of Izuu in Gion Shinchi, and the herring soba of Matsuba, next to the Minamiza theater.

2.4.2 The remaining signboards

In what was once a *sekigashi* table rental district surrounding Yasui Konpiragū Shrine, near the Kaburenjō in Gion Shinchi Kōbu, a signboard for the closed restaurant

Azumaya can still be seen (Fig. 2.7), announcing that "eel" *kabayaki* [dipped in sauce and grilled] was served in the "Edo style." The next chapter "Edo or Kansai?" will trace the history of *kabayaki* in Kyoto.

Figure 2.7 A signboard for Azumaya

As illustrated above, there were not only unique specialty restaurants in modern Kyoto that served specific fish and shellfish products but also restaurants that served a combination of different freshwater and saltwater fishes. The Azumaya signboard advertises "freshwater and saltwater fish dishes," indicating that it was the latter kind of restaurant. As one reads Kyoto (city) guidebooks from the Meiji period (beginning 1868) onward, it becomes clear that saltwater and freshwater fish dishes were already established as a genre therein.

The table below details the "Riuriya (ryōriya)" (Japanese-style restaurants) listed in the *List of Famous Places in Kyoto* (1877). Each restaurant is labeled with its genre along with its location, which makes it possible to understand the perceptions that shaped restaurant classification in the early Meiji era. The first item on the list is the long-established ryokan Hiiragiya, which also appeared in the introduction to this book. Following that, we have Nakamuraya (later renamed Nakamurarō) and Jiyū-tei, pioneers in the Westernization of Kyoto tourism, located on opposite sides in front of the torii gate of Yasaka Shrine, followed by Toganoo and Toriimoto, which are also located in front of the torii gate. The rest of the table includes iconic restaurants of modern Kyoto, such as Hyōtei and Heihachi Jaya, previously mentioned in this chapter. It is interesting to note that the word "eel" appears under the entry for Daiichi, a soft-shell turtle restaurant. Heihachi Jaya in Yamabata and Sagamiya in Shimogamo, both located along the road leading to Wakasa, are labeled as "Wakasamono" ("Wakasa produce"), perhaps to indicate the provenance of the produce, with salted fish from Wakasa being such an indispensable ingredient in Kyoto. There are 10 "saltwater and freshwater fish" restaurants and 8 "kaiseki saltwater and freshwater fish" restaurants. In the book *A One-man Guide to Buying and Selling* published the following year, in 1878, the genres of Kyoto cuisine are clearly organized into the following categories: kaiseki cuisine; *ikesu* freshwater and saltwater fish cuisine; loach set meal cuisine; Western-style cuisine; *sekigashi* cuisine; *hirukashiwa* cuisine; *shōjin* cuisine [vegetarian]; Buddhist *shōjin* cuisine; and beef cuisine.[63]

Table 2.1 Restaurants (*riuriya/ryōriya*) in the *List of Famous Places in Kyoto* (1877)

Location	Restaurant Name	Genre
Northern Aneyakōji Street and Fuyachō Street	Hiiragiya	Kaiseki saltwater and freshwater fish
Niken Chaya, Gion	Nakamuraya	Fish, Western-style
Niken Chaya, Gion	Jiyū-tei	Fish, Western-style
Shimogawara, Gion	Tsuganoo	Kaiseki saltwater and freshwater fish
Shimogawara, Gion	Toriimoto	Kaiseki saltwater and freshwater fish
Pontochō	Takemuraya	Kaiseki saltwater and freshwater fish
Pontochō	Ikeyoshi	Freshwater fish
Southern Shijō Street and Tominokōji Street	Matsumuraya	Kaiseki saltwater and freshwater fish
Northern Sanjjō Street and Kiyamachi Street	Ikegame	Freshwater fish
Northern Oike Street and Fuyachō Street	Hasshin	Kaiseki saltwater and freshwater fish
Eastern Tominokōji Street and Oike Street	Matsukiyo	Kaiseki saltwater and freshwater fish
Eastern Sakaimachi Street and Aneyakōji Street	Yaokan	Kaiseki saltwater and freshwater fish
Northern Shijō Street and Nishinotōin Street	Matsuyoshi	Freshwater fish
Western Karasuma Street and Uontana	Ōkame	Saltwater and freshwater fish
Nawate Street	Minokichi	Freshwater fish and birds
Nawate Street	Minosa	Freshwater fish and birds
Gionchō	Genchō	Saltwater and freshwater fish
4 Chōme, Miyagawachō	Shin'ya	Saltwater and freshwater fish
Northern Gojō Street, Miyagawachō	Tsuhan	Saltwater and freshwater fish
Shimogawara	Kaikaidō	Saltwater and freshwater fish
Sannenzaka Street	Akebono	Saltwater and freshwater fish
Nanzenji Temple	Hyōtei	Kaiseki
Shiwakawa Bridge, Sanjō Street	Fujikame	Saltwater and freshwater fish

Table 2.1 *(Continued)*

Location	Restaurant Name	Genre
Keage, Sanjō Street	Yumiya	Saltwater and freshwater fish
Oiwake, Sanjō Street	Yamakame	Saltwater and freshwater fish
Shimogamo	Sagamiya	Freshwater fish, Wakasa produce
Kamigamo	Kiyoroya	Freshwater fish, Wakasa produce
Yamabana	Heihachi	Freshwater fish, Wakasa produce
Yamabana	Gozan	Freshwater fish, Wakasa produce
Kitano	Kiya	Saltwater and freshwater fish
Gobanchō	Daiichi	Eel, turtle
Tōji Temple	Tobaya	–
Shimabara Station Exit	Aramatsu	Freshwater and saltwater birds
Nishinokyō	Sanoshin	–
Nishinokyō	Maruman	–

Note: Restaurants located in the Fushimi and Uji areas have been excluded. In the column "Location," where two street names are given, the location is the intersection of the two streets.

We have already discussed *sekigashi* cuisine; please see the relevant publication.[64] *Hirukashiwa* cuisine will be discussed in Chapter 4. Later lists added sushi and noodles as genres, and this categorization was generally followed in similar documents until World War II, albeit without "Buddhist *shōjin* cuisine." For example, the genres used in Daikyōto, the most systematic guidebook in the early Shōwa period were: general freshwater and saltwater fish cuisine; *imobō* cuisine; *shōjin* cuisine; soft-shelled turtle; oyster cuisine; eel cuisine; tempura cuisine; Western-style cuisine; Chinese-style cuisine; meat *sukiyaki* cuisine; chicken *sukiyaki* cuisine; sushi; *ochazuke* (tea rice soup); *Kanto-daki* (*oden* simmered items) (standing buffet); clay pot cuisine (one pot); and tasteful soba restaurants.[65]

This list of genres indicates a major change that occurred after the Meiji period: the emergence of *izakaya* (Japanese-style bars) where customers could order individual dishes such as *Kanto-daki* as and when they wanted them. The number of Western and Chinese restaurants was also steadily increasing. "General freshwater and saltwater fish cuisine" including kaiseki remained the major genre with more than 50 establishments.

However, in present-day Kyoto, there are no longer any restaurants in the city that serve saltwater and freshwater fish cuisine, other than Uoichi in Imakumano, which specializes in conger eel cuisine. The genre of "saltwater and freshwater fish cuisine" seems to have disappeared from modern Kyoto cuisine after the war or it is no longer advertised on signboards at least. It seems that rather than explicitly referencing "saltwater and freshwater fish" in the name of the cuisine, a convenient term has now been invented to replace it: *kyō-ryōri* (Kyoto cuisine).

Notes

1. Matsumoto, Seichō, "Kyō no ryōri" (Kyoto cuisine) (*Kyoto*, No. 45, 1986), pp. 2–3. Quote from p. 3.
2. Regarding various teahouses located in famous places, see Masahiro Kato's *Kyō no kagai monogatari* (Tales from Kyoto's entertainment districts) (Kadokawa Sensho, 2009, Chapter 2), and for more details on Niken chaya in Kamiya River, see Kato, Masahiro, ed., *Modan Kyoto: "Yūraku" no kūkan bunka shi* (Modern Kyoto: A spatial cultural history of "Yuraku") (Nakanishiya Publishing, 2017, Chapter 11).
3. Shiba, Kōkan, "Kōkan Seiyū Nikki" (Kōkan's diary of a journey to the West), (in Toshiro Koma et al. eds., *Shiryō: Kyoto kenbun ki, dai 2 kan, kikō II* (Historical materials: Record of Kyoto observations, vol. 2, travelogue II), Hōzōkan, 1991), pp. 264–269. Quote is from page 267.
4. Jippensha, Ikkū, *Tōkaidōchū Hizakurige, ge* (Shank's mare, vol. 2), Iwanami Bunko, 1973, Part 7, p. 267.
5. Kyokutei, Bakin, "Jinjutsu Kiryomanroku" (Tsukamoto Tetsuzō, ed., *Nikki kikō shū* (Compilation of travelogues), Yūhōdō Shoten, 1927), pp. 489–680. Quote from p. 574.
6. Tanizaki, Junichirō, "Suzaku Nikki (19)" (Suzaku diary, vol. 19) (*Osaka Mainichi Shimbun*, May 28, 1912).
7. Tanizaki, "Suzaku Nikki (16)" (May 25, 1912).
8. Kyokutei, "Jinjutsu Kiryomanroku", pp. 586–587.
9. It may have been Bakin's literary history that led to Shōtarō Ikenami's discovery of "udon" in Kyoto. Shōtarō Ikenami, *Mukashi no aji* (A taste of the past), Shinchō Bunko, 2003, p. 168.
10. Tanizaki, "Suzaku Nikki (19)."
11. Tanizaki, "Suzaku Nikki (7)" (May 4, 1912).
12. Tanizaki, "Suzaku Nikki (7)."
13. Tanizaki, Junichirō, "Sasame yuki, Jyō kan" (The Makioka sisters, vol. 1), *Tanizaki Junichirō zenshu, dai 15 kan* (The complete works of Junichirō Tanizaki vol. 15), Chūōkōronsha, 1982), p. 139.
14. Usui, Kinosuke, *Kyoto mikaku sanpo* (A gustatory stroll through Kyoto), Shirakawa Shoin, 1962, p. 26.
15. See "About Matsuba" (http://www.sobamatsuba.co.jp/about/index.html) on the Matsuba website. Last viewed on February 13, 2021.
16. *Rakumi*, vol. 2, no. 3, 1931, p. 72.
17. *Rakumi*, p. 72.
18. Note that Ai Tabata, the proprietress of Kyōsenryō, was the model for the heroine (Maiko Michitose) in Kyoshi Takahama's novels *Fūryū Zanpō*, *Zoku Fūryū Zanpō*, and *Fūryū Zanpō Gōjitsutan*. For more on this, see Kato, *Modan Kyoto*, Chapter 9.
19. Usui, *Kyoto mikaku sanpo*, p. 154.
20. Shimada, Fukuo, "Kyō hyaku aji: Sono 24" (One hundred flavors of Kyoto, Part 24) (*Rakumi*, vol. 75, 1958), pp. 102–104. Quote from p. 102.
21. Inoue, Jinnosuke, "Rakumi hanjōki (1) – Kyō no tabemono, natsu no maki" (Rakumi prosperity chronicle (1): Food in Kyoto Summer vol.) (*Rakumi* vol. 69, 1957), pp. 54–57. Quote from p. 55.

22. Usui, *Kyoto mikaku sanpo*, p. 155.
23. Kokubu, Ayako, *Kyoto Ajishirube*, Kasandō, 1980 (new edition), pp. 106–107. http://hibanaya.jp/ (last viewed on February 16, 2021).
24. Kunikida, Doppo (Compiled by Doppo Kunikida Editing Committee) *Kunikida Doppo zenshū* dai 9 kan (Complete works of Doppo Kunikida vol. 9), Gakken Kenkyūsha, 1966, p. 695.
25. "Ryōri ya Yaomasa, Asai Yasujirō shi" (Restaurant Yaomasa, Mr. Yasujirō Asai) (Onō Kōzō, ed., *Seikō Kamekan*, Osaka Jitsugyō Kōshinsho, 1909), pp. 126–128.
26. Shiga, Naoya, *Anya kōro* (A dark night's passage), Shincho Bunko, 2008, p. 366.
27. Makimura, Fumiaki, ed., *Osaka kotoba jiten* (The dictionary of the Osaka dialect), Kodansha Gakujutsu Bunko, 1984, pp. 158–159.
28. Yoda, Yoshikata, "Kamigata mikaku chizu 1: Ōtō" (Kamigata taste map 1: Ōtō) (*Tankō*, vol. 14, no. 1, 1960), pp. 66–69. Quote from p. 66.
29. From an advertisement at the end of *Rakumi* (vol. 192, 1968).
30. This restaurant was described as a branch of "Hiroshima Kakiharu, famous for its oysters" (Nishimura Zenshichirō ed., *Daikyōto*, Daikyotosha, 1928, p. 74). Newspaper advertisements from the same period referred to the restaurant on northern Kawabata Shijō as "Kakiharu Honten" (Main Restaurant) and that on the western end of Sanjō Ōhashi Bridge (near Kaburenjō) as "Kakiharu Shiten" (Branch Restaurant) (*Kyoto Nichinichi Shimbun*, October 31, 1927).
31. Inoue, Jinnosuke, "Rakumi hanjōki – Kyō no tabemono, sono 2" (Rakumi prosperity chronicle: Food in Kyoto, Part 2) (*Rakumi* vol. 70, 1957), pp. 62–65. Quote from p. 63.
32. Nishimura, *Daikyōto*, p. 104.
33. Kitaōji, Rosanjin, *Shun ka shu to: Ryori taikoku* (Spring, Summer, Fall, Winter: A great culinary nation), Chikuma Bunko, 2010, p. 213.
34. Kitaōji, *Shun ka shu to*, pp. 212–213.
35. Dōmoto, Kansei, "Meiji no koro no Kyoto ryōri ya" (Kyoto restaurants in the Meiji era) (*Rakumi*, vol. 100, 1960), pp. 34–36. Quote from p. 35.
36. Ishida, Kyokuzan, ed., *Kyoto meisho annai zue, furoku* (Illustrated guide to famous places in Kyoto, appendix), Shōhōdō, 1887, p. 47.
37. *Rakumi* (vol. 192, 1968) contains an advertisement for Heihachi Honten with the headline "famous grated yam, famous freshwater fish cuisine."
38. Kokubu, *Kyoto Ajishirube*, p. 100.
39. Natsume, Kinnosuke, *Sōseki zenshū, dai 19 kan* (The complete works of Sōseki, vol. 19), Iwanami Shoten, 1995, p. 298.
40. Usui, Kinosuke, *Sinpen: Kyoto mikaku sanpo – Kyō no gaido* (New vol.: A gustatory stroll through Kyoto – A guide to Kyoto), Shirakawa Shoin, 1970, p. 399.
41. Ōkubo, Tsuneji, "Kamigata mikaku chizu 6: Yase Ōhara wa sato no aji" (Kamigata taste map (6) Yase Ōhara – Taste of the village) (*Tankō*, vol. 14, no. 6, 1960), pp. 44–46. Quote is from p. 46.
42. Yanagita, Kunio, *Meiji Taishō shi, sesō hen, shinsō ban* (Meiji Taishō history, social conditions, new edition), Kodansha Gakubunko, 2000, p. 77.
43. Endō, Mōhei, ed., *Kyoto meisho annai zue* (Illustrated guide to famous places of Kyoto), Fukui Genjirō, 1881, p. 20.

44. *Rakumi*, vol. 75, 1958, p. 101.
45. Kitaōji, *Shun ka shu to*, p. 237.
46. *Rakumi*, No. 2, 1935, p. 19.
47. Umezaki, Shujaku, "Dojō ryōri" (Loach cuisine) (*Rakumi*, vol. 65, 1957), pp. 90–91.
48. Inoue, Jinnosuke, "Rakumi hanjōki – Kyō no tabemono, sono 7" (Rakumi prosperity chronicles: Food in Kyoto part 7) (*Rakumi* vol. 75, 1958), pp. 98–101. Quote from p. 100.
49. Hawley, Frank, "Suppon ryōri" (Soft-shelled turtle cuisine) (*Kaizō*, vol. 17, no. 3, 1935), pp. 293–299. The subsequent quotes are also from this essay.
50. Saeki, Teiji, "Tanmi henro: Daiichi suppon" (Tanmi pilgrimage: Daiichi soft-shelled turtle) (*Rakumi*, vol. 1, no. 2, 1935), pp. 30–34. Quote from p. 32.
51. Kitaōji, *Shun ka shu to*, pp. 96–97.
52. Kitaōji, p. 110.
53. The Hiranoya in Toriimoto, which will be discussed later, is described as gathering "sweetfish from the upper reaches of the Hozu River, Utsu, Wachi, Seki, Shuzan, etc." in the "high season" and then as "selecting the best ones and delivering them to prestigious restaurants in Kyoto, such as Hyōtei, until the beginning of the Taishō period." In doing this, "life flapping sweetfish were carried around in water in a bucket, which was sorted using scales." Kokubu Ayako and Nishiyama Jirō, *Kyō no okami tachi: Shinise no aji o tazunete* (Kyoto hostesses: Visiting the long-established restaurant), (Shibata Shoten, 1980), pp. 156–157.
54. Inoue, "Rakumi hanjōki, sono 7," pp. 98–99.
55. Inoue, p. 99.
56. Mizukami, Tsutomu, *Shi han Kyoto zue* (Private edition of Kyoto illustrations), Fukutake Bunko, 1986, p. 174.
57. Ikenami, Shōtarō, *Sanpo no toki nani ka tabetaku natte* (When I take a walk, I want to eat something), Shinchō Bunko, 1981, p. 213.
58. Ikenami, Shōtarō, *Edo no aji o tabetaku natte* (I want to taste the flavor of Edo), Shinchō Bunko, 2010, p. 52.
59. *Rakumi* No. 2, 1935.
60. Yamakawa, Mikumi, "Aji to kibun o tazunete (2)" (Visiting taste and mood (2)) (*Rakumi*, vol. 1, no. 2, 1935), pp. 75–82. Quote from p. 77.
61. Usui, Kinosuke, *Kyoto mikaku sanpo* (A gustatory stroll through Kyoto), Shirakawa Shoin, 1963 (5th edition), p. 214.
62. Usui, *Kyoto mikaku sanpo* (5th edition), p. 206.
63. Harada, Yosamatsu, *Bai bai hitori annai* (A one-man guide to buying and selling), Yosamatsu Harada, 1878.
64. Kato, Masahiro, *Kyō no kagai monogatari* (Tales from Kyoto's entertainment districts), Kadokawa Senshu, 2009, Chapter 3.
65. Nishimura, *Daikyōto*, pp. 74–80.

10.2478/9784991351600-010

Chapter 3
Edo or Kansai?

> Looking at eel restaurants in Kyoto today, many of the famous establishments – with some prominent exceptions, such as Izumoya, Sanmanju, Kaneyo, and Maisaka – use the phrase “Edo-yaki” on their signboards.[1]

In the Kitano shopping district close to Kitano Tenmangū Shrine, there is a market-style food court called Shinagawa Food Pool. At the entrance of this food court is Kamata Kawaoten, a long-standing restaurant that specializes in eels, with a signboard that reads, “Kansai-style eel *kabayaki*.”[2] If one arrives too late, there will be a small wooden sign that reads, “Kansai-style eel *kabayaki* sold out today.”

This phrase “Kansai-style,” more generally called “Osaka-style,” refers to the *jiyaki* method, in which the eel is first slit open along the abdomen, and then placed directly on the grill. This method contradicts *Edo-yaki* (the so-called Edo/Tokyo style) in which the eel is slit open along its back and is first grilled *shira-yaki* (without any seasoning), gently steamed, and then seared again at the end. The writer Nozomi Hayashi called the Kansai-style *kabayaki* eel “crunchy” because of its texture.[3]

In Kyoto today, as per the quote at the beginning of this chapter (although it was written about 60 years ago now), *kabayaki* served at eel restaurants is generally Edo-yaki. Tracing the origins of Edo-yaki in Kyoto, contrary to some claims that “Edo-style *kabayaki* started to feature on restaurant signboards in Kyoto at the end of the Edo period,”[4] the restaurant Kandagawa, formerly located in the Saiseki area of Kyoto, in fact, gets the credit for this part of Kyoto’s modern history.

3.1 The Edo-yaki of Kyoto: Kandagawa

3.1.1 Tales of provenance

According to Kinosuke Usui:

> When one speaks of long-established restaurants offering Edo-yaki, the first place that comes to mind is Kandagawa in Saiseki.
>
> In 1884, Mr. Shigekata Miyata (the current proprietor’s grandfather), who had previously been a district headman in Gunma Prefecture, opted to run a 3,000-*tsubo* [just under a

> hectare] hot spring inn and [a restaurant called] Shinsentei in Nezu-suga-chō, Hongō, Tokyo, having no wish to become the governor of Niigata Prefecture. After talking with the owner of Kandagawa in Kanda Myōjinshita, Mr. Miyata started an eel restaurant at Shinsentei. The Kyoto Exposition was held in 1894, and at the recommendation of Hiroshi Nakai (Ōshū), who was the governor of Kyoto Prefecture at the time, Mr. Miyata decided to establish an Edo-style eel restaurant to the east of the Minamiza Theater on Shijō as there were no restaurants offering this cuisine in Kyoto at the time. Thereafter, having been well received by Kyoto people, the restaurant remained open even after the end of the exposition and moved to its current location in the following year, 1895, where it remains today.[5]

With such a history, the Kandagawa in Saiseki became positioned as "the pioneer of Edo-yaki eel in Kyoto." In the early 1930s, the following article was published, expanding on the second paragraph in the above quote:

> As this kind of cuisine did not exist in Kyoto at the time, on the advice of the honorable Governor Hiroshi Nakai, the late Mr. Shigekata Miyata opened a branch restaurant to the east of Shijō Minamiza in 1894 and started serving pure Tokyo-style eel cuisine under the name of Kandagawa.
>
> Thereafter, having been met with enthusiasm by the people, the restaurant moved to its current location in the subsequent year, 1895, and has seen ever more success to this day... Mr. Shinzō Ōtake, the third-generation owner of the restaurant, continues to make efforts – of which people are largely unaware – to use only wild eels and not use any farmed eels whatsoever, so as not to sully the *noren* curtains at the restaurant's entryway [meaning "keep the restaurant's good reputation untarnished"].[6]

Other similar narratives include the following:

> The third-generation owner of this establishment was originally from Gunma Prefecture. In 1885, having been the headman of Yamada County in the same prefecture, he was specially selected to be the governor of Niigata Prefecture. However, having no wish to become governor, he went to Tokyo and started a ryokan and eel restaurant called Kandagawa in Nezu-suga-chō (with the main Kandagawa restaurant named Kanda Myōjinshita).
>
> Ten years later, in 1895, when the Exposition was held in Kyoto, on the recommendation of his friend, Kyoto Governor Hiroshi Nakai, he opened a Kandagawa restaurant next to the Minamiza Theater in Kyoto. The restaurant then moved to its current location in the following year and began operating in earnest.
>
> It is for this reason that the restaurant's advertising slogan is "Seventy years of Edo-yaki in Kyoto." The current owner, Yasuo Ōtake, is the fourth generation. In other words, Kandagawa is the pioneer of Edo-style eel in Kyoto.[7]

We can combine and summarize the episodes described in the two guidebooks as follows. First, Kandagawa (in Saiseki), the pioneer of Edo-yaki in Kyoto, originated from Myōjinshita Kandagawa, a famous eel restaurant still in business today.

The first-generation owner of the Saiseki Kandagawa was Shigekata Miyata, who had served as a district headman in Gunma Prefecture. In 1884, he declined a promotion to the post of governor of Niigata Prefecture, and in the following year, 1885, he instead sought a large plot of land in Nezu-suga-cho, Hongō, Tokyo, and "operated a hot spring inn and [a restaurant called] Shinsentei." Although the connection between Miyata and the owner of the Kandagawa in Myōjinshita is unclear, it seems that he either changed Shinsentei into an eel specialty restaurant or that he opened it as a *kabayaki* restaurant.

Although some important information can be extracted from these narratives, their hearsay-like nature leaves many points of ambiguity. What we will focus on here is the location of Shinsentei (Nezu-suga-chō). The name "Nezu" in the context of Meiji Tokyo suggests a red-light district that was in front of the gate of Nezu Shrine, said to be the forerunner of Suzaki, which later developed into a very popular red-light district in the imperial capital alongside Shin-Yoshiwara.

Delving into the history of Nezu reveals an unexpected connection between Gunma, Tokyo, and Kyoto.

3.1.2 Relocation of the Nezu red-light district

Jirō Matsukawa's *Nationwide Tour of Entertainment Districts* (1929) is the most systematic guidebook on *kagai* entertainment districts (red-light districts) in the prewar Shōwa period. It includes the following passage in the section on Suzaki:

> The Nezu brothel was established in 1868 and was located in what is now Nezu-yaegaki-chō and Suga-chō until that time [when it was relocated to Suzaki in 1888], such that even today, many charming names remain in that vicinity, such as the Mikaeri [Looking Back] Bridge, Tedori [Hand-Holding] Bridge, Tasogare [Twilight] Bridge, and Aizen (Aizome) River [Indigo Dye].[8]

Having developed as an *okabasho* (another term for a brothel district), Nezu survived through the turmoil of the last days of the Tokugawa shogunate and the Meiji Restoration, but its closure was planned in 1884 to make room for Tokyo Imperial University, which was to be situated "on Mukogaoka, which is connected to Tsutsujigaoka and Nishioka directly above." In 1888, the main brothels were relocated to the new red-light district Suzaki on reclaimed land.[9]

The key point here is the relocation Ōyawatarō, which was the largest brothel in Nezu:

> Yawatarō, a brothel in Nezu, shall be removed from its location at the present time, and *its current building shall be left on the site as a large hot spring*. Another new building shall be built on the reclaimed land in Suzaki, and all the prostitutes who have been running the business up to this time have been summoned there. Those among them who wish

> to migrate to Suzaki will be taken to the new building, and those who are perhaps a little too old and no longer wish to work there shall be permitted to discontinue their business and transfer to working as maids at the hot spring. This arrangement was satisfactory to all, with each thinking about which way to go.[10]

Ōyawatarō ("Yawatarō" in the text) was located within the precincts of the Nezu Shrine, on both sides of the Nezu district that formed the brothel area. This leads us to the conclusion that the "large hot spring" mentioned in the quote is the one that would later become Shinsentei.

Shinsentei was originally a mineral hot spring that opened in 1884, located in Uguisudani, close to Nezu Shrine.

> Isobe Hot Springs: Having divined a suitable site with deep tranquility and seclusion, in Uguisudani, beneath Shinzaka in Ueno, a salt spring with carbonic acids was opened the day before yesterday in Isobe, Usui District, Jōshū.[11]

Isobe Hot Springs was so named because spring water was brought in from mineral hot springs in Isobe in Jōshū via the Nakasendō highway.[12] When Nezu relocated to Suzaki, this hot spring facility – that is, Shinsentei – was converted from the vacant former brothel Ōyawatarō.

Here, let us quote at length from the "Relocation and Opening Advertisement:"

> Thanks to your overwhelming patronage, our humble establishment has thrived; however, with our limited capacity, we regrettably had to turn away bathers from our temporary venue on a daily basis. *As part of our planned renovation, we have taken over the vacant property of the Ōyawatarō Manor in the Nezu entertainment quarter.* The establishment will relocate and reopen at this new location on the 6th of the month, offering fresh baths every day. We sincerely hope for your continued or even redoubled support. The new location is an elegant place, boasting tall trees, an elevated and dry atmosphere with clean air, and potable spring water that ranks among the best in the prefecture. Since ancient times, it has been renowned as a place for the convalescence of beriberi sufferers. Merely two blocks from Taitō, one can visit the chrysanthemums of Dangozaka, the lotuses of Shinobazu Pond, and Shiitake Mountain – a location of choice for mushroom foraging – all in close proximity and ideal for seasonal sightseeing. Furthermore, to provide leisure and recreation to our clientele, we have newly established a large archery range, swing set, fishing pond, and – for Go enthusiasts – a branch of the Hōensha. What was only yesterday a bustling and libertine pleasure quarter has today been transformed into a haven of tranquility and recreation. Bathing in the efficacious waters of the sacred spring and breathing in the fresh air will no doubt contribute to the health of the entire prefecture. However, all prices will remain the same...
>
> Exclusively available at the Shinsentei Teahouse, located at the Isobe Hot Springs Lodge in Nezu-Tsutsujioka, Tokyo, a Top-Tier Seiseisha Hot Spring Resort.[13]

The Shinsentei Teahouse at the Isobe Hot Spring in Nezu opened on October 6, 1888.[14] Let's review this new Shinsentei and its environs as they were described in this advertisement.

As mentioned, the Shinsentei was established by repurposing the Ōyawatarō, a large residence that had been left vacant after the relocation of the Nezu red-light district. Nezu is located on a raised plateau, which is seen from an eastern vantage such as Negishi, and was long regarded as "a high and dry area with clean and crisp air and good water drainage."

Dangōzaka and Shinobazu Pond, respectively famous for their dolls made of chrysanthemums and their lotus flowers, were both located nearby and were suitable for sightseeing. Besides this, "a large archery range, a swing set, a fishing pond, and a branch of Hōensha" were provided for the comfort and entertainment of users. The site usage seems to anticipate the amusement parks that were developed in the suburbs throughout the modern era.

The advertiser's designation as "top-tier" (*kōgō-sen*) refers to the type of mineral spring, and "Saiseisha" is the name of a limited partnership company located in Isobe Kōsen (Isobe Hot Springs, Annaka City, Gunma Prefecture) that managed all aspects of the business, including the sale of mineral springwater.[15] As suggested by the name Tsusujigaoka (literally Azalea Hill), azaleas are still cultivated on a slope on the west of the Nezu Shrine precincts. The following year, an advertisement read, "Those who want to spend the summer in a quiet place in the prefecture should go to Shinsentei in Tsutsujigaoka, Nezu. Located on the site of the former Ōyawatarō, it is a convenient place to cool off while enjoying the cuisine and spacious gardens of Isobe Hot Springs."[16] Thus, Shinsentei was initially operated as a hot spring ryokan that also offered food.

The month after this advertisement was published, an interesting article appeared in a separate newspaper:

> The Economy in Nezu
>
> After the pleasure district moved to Suzaki last year, the site of the pleasure district in Nezu suddenly became quiet due to the recent modifications to the city districts. The economy has recovered slightly, especially with the summer heat, and more than 200 people visit Isobe Hot Springs after Ōyawatarō every day. *The Kiyomizutaki eel restaurant, named Kandagawa, located near the rear gate of Nezu Shrine, also has a considerable number of visitors.* Over the last two days, three archery ranges and one geisha house have opened for business, and there are others who plan to open other establishments, such as table service restaurants. Therefore, it seems assured that the area shall be restored to its former prosperity in due course.[17]

Although there were numerous cases wherein former red-light districts were revived as pure *kagai* entertainment districts with only geisha,[18] in this case, the eel restaurant Kandagawa featured in Nezu, which was planned to be revived as an amusement park on the outskirts of the imperial capital. Below is another advertisement from 1891 with the headline "Shinsentei, Nezu, the fairyland of Tokyo":

> In the dusty environs of the imperial capital, a separate banquet district has been formed. It is our pavilion [i.e., restaurant] alone that can paint a picture of a truly enchanted land. The pavilion is located in front of the beautiful offerings of the Kanda River [Kandagawa], surrounded by clear ponds and green mountains. Within the sacred baths of the eight salt springs of Isobe, the pavilion awaits the visit of distinguished guests.
>
> More than 50 guest rooms
>
> A 130-tatami mat hall[19]

In September of the previous year, 1890, Shinsentei was extensively renovated, including the installation of a hall with a total floor space of 130 tatami mats.[20] These two articles indicate that there was an eel *kabayaki* specialty restaurant called Kandagawa, separate from Shinsentei's main building, which was a hot spring facility converted from the former Ōyawatarō.

3.1.3 Shigekata Miyata and Isobe Hot Springs

When we look at the accounts of the history of Kandagawa's expansion into Kyoto, we realize that they contain points of uncertainty and obvious errors.

Shinsentei opened in Uguisudani in 1884. Both accounts quoted in Section 3.1.1 refer to Nezu-suga-chō. The date of 1885 given in the latter source is likely an error. Both sources describe Shigekata Miyata as "having no wish to become" the governor of Niigata Prefecture.

Going through the *Kan'inroku*[21] (the list of civil servants [government officials] of the Meiji government) reveals that Shigekata Miyata is indeed listed in a section on Gunma Prefecture, although there are some differences with the above narratives. He held the position of "Second-Rank Police Inspector" until July 1878, and in August of the same year, he was promoted to the most senior rank below *kenrei* (equivalent to the modern-day prefectural governor [*kenchji* in Japanese]) and Chief Secretary. The characters for "Ehime" are written above his name, so it appears that he was from Ehime Prefecture.

According to the *Kan'inroku* dated January 1879, Miyata became the "district headman" of "Kataoka District in western Gunma." In November 1880, the *Kan'inroku* lists him under that position; however, from December 1880, Miyata's name disappears from the lists of Gunma Prefecture government officials. In other words, the above accounts were correct in stating that he was a former "district headman," but there is no evidence to suggest that he was the "head of Yamada district."

That being so, the following hypotheses may also be valid. Under the Meiji government, Shigekata Miyata became a local government official in the police department, and after ascending to the third most important post in the prefecture, he was transferred to become the headman of Kataoka District, before retiring after about two years in office. Four years after that, in 1884, Miyata was approached for the position of governor of Niigata Prefecture, but for some reason, he declined and chose a different path; he instead chose to open Shinsentei (Isobe Hot Springs) in Uguisudani. Following the sequence of events this way, we can see part of the reason why Shigekata Miyata was able to open Shinsentei. Although born in Ehime Prefecture, Miyata had worked as a local bureaucrat in Gunma Prefecture for the Meiji government, where he built a career before retiring, sometime after which there was talk of him becoming governor of Niigata Prefecture. However, rather than returning to his government post, Miyata chose to spend the rest of his life as a manager. Having been involved in the administration of Gunma in the early Meiji period, he must have had a wide circle of acquaintances, which might have also included those involved (Saiseisha, for instance) with the newly developed mineral springs at Isobe Hot Springs. Then, in July 1884, Shinsentei opened for business in Uguisudani.

Although the accounts of Miyata in Kyoto are muddled, it does seem that Kandagawa opened in Nezu-suga-chō around 1889. In other words, it appears that Miyata first opened Shinsentei in Uguisudani, then moved to Nezu, and then later opened a branch of Kandagawa serving eel cuisine.

3.1.4 The establishment of Kandagawa in Kyoto

To reiterate, in Section 3.1.1 we saw that the Kandagawa of Nezu expanded into Kyoto in 1894. The following newspaper advertisement confirms this:

> Advertisement for the Establishment of a Branch Restaurant in Nishikyo [Western Kyoto]
>
> In accordance with the recommendations of our honored customers, [Kandagawa] shall establish a branch restaurant in Shijō Saiseki, Kyoto. We hope that our customers will extend their patronage to this branch restaurant as they do with our main restaurant, on the occasion of the anniversary festival and exhibition.
>
> Tokyo Kabayaki, Kanda Myōjinshita, in front of Nezu Shrine, Kandagawa Honten/Shiten
>
> Triumphal Return Stew, Takeover Soup, Tokyo Nezu Shinsentei[22]

This advertisement ran in the *Yomiuri Shimbun* in 1894. The menu items "triumphal return" and "takeover" express the nationalistic jingoism that was rife following the outbreak of war between Japan and China. The Advertisement clearly states that the Shinsentei *kabayaki* restaurant in Nezu was a branch of the original Kanda Myōjinshita Kandagawa. The phrase, "[Kandagawa] shall establish a branch restaurant in the Shijō Saiseki area of Kyoto ... on the occasion of the anniversary festival exposition" is also worth noting, as it provides information about when and where the new branch opened, although this information may need some unpacking.

Let us start by looking at the second half of this phrase. "Anniversary festival" (*ki'nensai*) and "exhibition" (*hakurankai*) respectively refer to the 1,000-year anniversary of the transfer of the capital to Heian-kyō (modern-day Kyoto) and the Fourth National Industrial Exhibition. Kandagawa established premises in Kyoto in November 1894, apparently at the invitation of the governor of Kyoto Prefecture, Hiroshi Nakai (Ōshū). This is quite possibly true as his term of office was from November 1893 to October 1894. However, the exhibition was held from April 1 to July 31, 1895, with the anniversary festival scheduled for April 30 (although it was actually postponed to October), so the accounts of the provenance of the restaurant given in the two guidebooks above are incorrect, as is the other narrative that "Miyata opened a restaurant to the east of Shijō Minamiza ... and moved to its current location in the subsequent year, 1895."

Because Japanese does not usually distinguish between singular and plural, the newspaper advertisement's phrasing, "branch restaurant in Shijō Saiseki, Kyoto" can also be read as meaning that branch restaurants were established in both the Shijō and Saiseki areas, in which case, the former would have been next to the Minamiza Theater. At least at this stage, there is no doubt that a restaurant was opened in Saiseki. Indeed, the guidebook *Guide to the One Hundred Faces of Kyoto*, published in December 1894, contains the following entry: "Tokyo *kabayaki*, Kandagawa, Saiseki, north Shijō."[23] Even if they had been operating in two locations, it seems likely that the business consolidated in Saiseki after the end of the National Exhibition, where it became permanently established in Kyoto.

Kandagawa was the originator of Edo-yaki eel in modern Kyoto, and Kandagawa itself had its origins in Nezu. It is unclear what became of Shinsentei in Nezu thereafter.

3.2 Tokyo connections

3.2.1 Umenoi in Shinbashi

In the history of Edo-yaki in Kyoto, another important restaurant is the previously mentioned Gion Umenoi, which is now located in Kitabayashi, Maruyama Park. This restaurant also has its roots in *kabayaki* restaurants of Tokyo, as we can see in the following account: "after returning from training at Umenoi in Shinbashi, Tokyo, in 1919, the founder opened a restaurant with the same name, with 'Gion' added in front of it."[24] At the beginning of the Shōwa era in 1926, there were three branches of Umenoi in Tokyo;[25] but the reference to "Shinbashi" means that he trained at the restaurant located at 1-chōme, Shibaguchi-chō, Shiba Ward (currently 1-chōme, Shinbashi, Minato Ward).

The Umenoi in Shibaguchi-chō seems to have been in business since the early Meiji period (1868), and it seems that the founder of Gion Umenoi was trained at the restaurant owned by Toyokichi Hayashida, who opened a "*kabayaki* specialty" restaurant in June 1896, taking over the premises and name of an existing restaurant.[26] This restaurant was remodeled and renamed "Shinbashi Umenoi" to coincide with revisions to the city districts, and it seems that the business flourished, with a new branch opening in the Tamachi area of Akasaka in Tokyo and a large expansions to the main restaurant.

Incidentally, like the quote above, postwar guidebooks all state that Gion Umenoi was founded in 1919, but Umenoi's own website claims that it was established in 1914.[27] Moreover, although the sources do not confirm whether the Kyoto branch was prefixed with "Gion" from the beginning, the following explanation may shed some light on the matter:

> The founder honed his craft at the Umenoi in Shinbashi, Tokyo. He initially opened a restaurant near the Minamiza Theater, where his flavors earned praise from theater goers, gaining him a good reputation. The restaurant then moved to its current location in the Gion district, but the customers in this area are often hard to please, so it is said that he puts even more effort into his cooking and food preparation so as not to lose his good name.[28]

Thus, it appears that Umenoi initially opened near Minamiza in 1914, where it received acclaim, before moving in 1918 to "its current location in the Gion district," the north of the intersection between Yamatoōji and Shijō roads (i.e., Nawate-dōri). Note that this location is not actually within the Gion district; rather, the restaurant added "Gion" to its name because it was located on Nawate-dōri, which is on the edge of the Gion Shinchi area. There are no other examples of long-standing establishments on Nawate-dōri that include "Gion" in their name, so it would be rather unusual if the "Gion" prefix had been included from the beginning, and indeed, none of the advertisements for Umenoi from the time mention Gion.

The above quote also states that Umenoi was located near Minamiza, which was also once the site of Kandagawa. Recalling that Kandagawa later moved to Saiseki in Pontochō on the other side of the river and Umenoi moved to the edge of Gion Shinmachi, it is clear that both selected locations close to *kagai* entertainment districts. It is probably not a coincidence that Myōjinshita Kandagawa was a destination for geisha with clients from the Kōbusho military academy in the Kanda area of Tokyo, and that Shinbashi Umenoi was located near the Shinbashi *kagai*.

3.2.2 The creation of a Kyoto specialty

Saiseki and the area between Sanjō and Shijō on Yamato-ōji street (i.e., Nawate-dōri) were the restaurant districts that typified early modern Kyoto. During the Meiji period, "on both sides of the Kamo River, there were several freshwater fish specialist restaurants in Pontochō and Nawate," including Minokichi, Minoshō, Uoshin, Chirirenge,[29] and the freshwater fish wholesaler Kanemasa. After the war, apart from Minokichi (whose main restaurant was called Takeshigerō), which moved to the Hiromichi area in front of Heian Jingū Shrine, there were no freshwater fish restaurants left on Nawate-dōri, and Kanemasa is the only establishment that continues its business with the air of a merchant family of old. The founder of Kanemasa was reportedly from Ōtsu in the neighboring Ōmi province and moved to the current location during the Keiō era (1865–1868): "At one time, they expanded their business not only in Kyoto but also into the Osaka area, supplying fish to Shibatō and Izumoya, renowned eel restaurants in Osaka."[30] Locals in the area still recount how Kanemasa used to moor a boat in the Lake Biwa Canal, which flowed along the western side of the restaurant (between present-day Kawabata-dōri), where they would cast a net and release freshwater fish into it.

What made Kanemasa famous as an eel restaurant rather than simply as a general wholesaler of freshwater fish was the commercialization of *ochazuke-unagi* (green tea rice soup with eel) by the second-generation owner. "The reason for developing this dish was said to be to make effective use of eels that were a little too much on the small side to be made into *kabayaki*."[31]

However, there is another more interesting story about the dish's origins:

> The previous owner [of Kanemasa] had become friends with various individuals through their mutual love of eel, including the writer Murakami Namiroku, the owner of a long-established eel restaurant in Tokyo, and the owner of Ōwada in Roppongi. When the previous owner of Kanemasa heard that Kyoto had no specialty dish that centered on eel, he was inspired to try creating various potential eel dishes with his wife. Eventually they came up with *ochazuke-unagi*, which they gave to the three eel aficionado friends named above, all of whom told them it was good; so, they started selling it to the public.
>
> The dish is made by taking an eel around 40 centimeters in length, known as a *biri*, preparing it in the Kanto style by slitting it open along the back and then grilling it without seasoning, before simmering it for hours with mirin and soy sauce. The combination of several kinds of soy sauce gives it an indescribable flavor. It may seem strange that a fatty food such as eel would go well with *ochazuke* [tea rice soup], but it does indeed taste delicious if you make the *ochazuke* with a high-quality sencha tea.[32]

Thus, the second-generation owner apparently created the specialty *ochazuke-unagi* at the provocation of his "eel aficionado friends". Thus, Kyoto's connection with Tokyo eel is deeper than one might initially expect.

3.3 *Kabayaki* rivalry

> In Kyoto, the *Edo-yaki* restaurants are Umenoi (Nawate) and Kandagawa, and Minokichi (Nawate) and Minoshō (in front of the Imperial Palace) also serve eel.[33]

3.3.1 Neither Tokyo-style nor Kansai-style

As per the quote at the beginning of this chapter ("many of the famous establishments – with some prominent exceptions, such as Izumoya, Sanmanju, Kaneyo, and Maisaka – all use the phrase 'Edo-yaki' on their signboards")[34], there is a sense that Tokyo-style *kabayaki*, with its roots in Kandagawa and Umenoi, became firmly established in Kyoto.

Looking through the telephone directory from 1926 to 1939, both the main restaurant Umenoi Honten (Nawate-dōri) and the branch Umenoi Shiten (north of Yanagibaba Anekōji) are listed as "Tokyo-style eel *kabayaki* cuisine," whereas the branch Izumoya Maruyama Shiten (Maruyama Park, Kitabayashi) is listed as "Tokyo-style eel freshwater fish cuisine."[35] Izumoya is also an Edo-yaki restaurant that still has premises in Pontochō,[36] as is Kaneyo, which continues to operate on the outskirts of the Shinkyōgoku area. Given that some restaurants have changed the genre of their cuisine in this way, it may be more appropriate to say that Edo-yaki has become the norm, rather than an illusive specialty in Kyoto.

However, two specialty restaurants trod their own paths in postwar Kyoto: Matsuno and Maisaka. Let us hear again from *kabayaki* lover Jinnosuke Inoue:

> In front of the Minamiza theater, the fourth establishment on the street as one is walking toward Gion has a sturdy frontage in a folk-art style. This is the eel restaurant Matsuno; to the left of the narrow entrance is the kitchen, where one can see the figures of chefs grilling eel as one passes by. And mingling with the people in the kitchen, dressed in the same white work clothes, one can sometimes see a tall woman moving about with tireless energy. Despite being the proprietress of the restaurant, I do not believe that there are many people who know that it was she who worked hard to build Matsuno up to what it is today with her slender arms.
>
> Matsuno was founded in 1940 and was originally a restaurant specializing in soba noodles, but as her younger brother was running the nearby Matsuba, a famous herring soba restaurant, she wondered if it would be a good idea to also carry on as a soba restaurant. She looked into various options before reopening as an eel restaurant in October 1953.[37]

Thus, as we can see, Matsuno has close connections with Matsuba, famous as the originator of *nishin soba* (herring soba, discussed above). Matsuno opened in 1940, and after Matsuba was forced to evacuate its premises to preserve the Minamiza Theater at the end of WWII, Matsuba moved and reopened at the same premises as Matsuno. When Matsuba reopened in a new four-story reinforced concrete building on its original site in 1951, Matsuno, which had for a brief spell been "a restaurant catering to those with a sweet tooth," relaunched as a restaurant specializing in eel dishes in the fall of 1953.[38]

After Jinnosuke Inoue died in 1973, Ayako Kokubu took over the Matsuno entry in the sequel to the *Kyoto Taste Map*. She notes that "Matsuno opened its premises with a self-proclaimed 'Matsuno-style,' which is neither Edo-yaki nor Kansai-style, but rather an accumulation of techniques to produce a good flavor."[39] In fact, Matsuno's current website states that it adds "steaming" to "Matsuno's unique approach to 'grilling,' which is not bound by the Edo or Kyoto styles."[40]

The postwar style *kabayaki* in Kyoto seems to have skillfully blended the cooking methods of Tokyo and Kansai. When Jinnosuke Inoue visited Matsuno to research its entry in the guidebook, he found that Matsuno "sourced its eels directly from Maisaka [referring to the place], from a single selection of good fish from Lake Hamana."[41] The kanji used in the place name Maisaka (舞阪) are different from those used in the name of the restaurant Maisaka (舞坂). However, the restaurant Maisaka was indeed named after the place, and like Matsuno, it opened its doors as an eel restaurant not long after the war in the Kiyamachi Rokkaku area of Kyoto.

The owner reportedly worked at a restaurant named Edogawa on Kayū-kōji street before establishing her own business, so she would have been familiar with Edo-yaki; however, when she opened her own restaurant, she created her own grilling method.

> One should refer to the eel here as "Kyō-yaki" [Kyoto-style] because it is slightly different from the general Kansai-yaki. Usually, the eel is first fully grilled without any seasoning and then only minimally grilled after the sauce has been applied, whereas at Maisaka, it is lightly grilled first, and then fully grilled after the sauce has been applied.[42]

Maisaka was located in Kiyamachi Rokkaku (or more precisely, Kita-kurumayachō on the right bank of the Takase River), in the center of Kiyamachi, which had become something of a nightlife district. Its opening hours were thus from three in the afternoon to three in the morning, which meant it wasn't open for lunch. So, it had a sign at the front which read: "We kindly request our lunchtime customers to visit our branches at Kawaramachi Rokkaku and the entrance of Daini Kyōgoku [aka Shinkyōgoku]. Please visit our main restaurant when you would like something to drink later in the evening." Here we see one of the defining features of the history of *kabayaki* in Kyoto: the postwar style of *kabayaki* in Kyoto was neither in the style of Tokyo nor Kansai, but instead had its origins in the "nightlife district" near Minamiza, where the first restaurants to serve Edo-yaki in Kyoto (Kandagawa and Umenoi) may have been located.

3.3.2 Eel as a last course

Let us now return to the quotation at the beginning of this section. We have already seen Kandagawa and Umenoi, but the fact that traditional restaurants in Kyoto such as Minokichi and Minoshō also "serve eel" means that eel was served at freshwater fish restaurants, rather than at *kabayaki* restaurants (eel specialty restaurants), as was typical in Tokyo. We learned in Chapter 2 that Minoshō was a loach specialty restaurant.

Looking into this further, we learn that a surprising number of well-known restaurants in Kyoto once served *kabayaki*. According to Doppo Kunikida's "Amateur Reviews of Cuisine in Western Kyoto:"[43]

> The cost of a meal is two yen per person, consisting of four dishes: the family's famous soft-shelled turtle soup, eel *kabayaki*, *iridori* [braised chicken], and sashimi.

This passage is from a review of Hyōtei, and it was certainly not all positive.

> "*Kabayaki* eels" are large and about five *sun* [approx. 15 cm] wide, so they tend to be excessively fatty and not the best in quality. Especially in Tokyo, the eel is cut open along the back, and the small fin bones are removed along with the backbone, so one may even be unaware that there were any bones. However, in this area, the eel is opened along the belly and only the backbone is removed, without removing the small fin bones so that when it comes to the larger eels, one is cognizant of the bones. [Therefore], when it comes to how the eel is cut, Tokyo's method is better. Another defect is that the skin of the eel is very tough, unlike it is in Tokyo. A companion from Tokyo criticized the dish, finding it too salty and saying that he could not bear to eat it. Although I did not entirely agree with him, I found the tough skin of the eel to be unpalatable, and thus, I ate only the flesh. If it would have been a medium-sized eel, such flaws would not have been evident. Because of this cooking method, the eel ordered in Kyoto and its environs should be medium-sized or smaller, and the larger ones are best avoided.[44]

Hyōtei retains its good reputation today, but it seems that the *kabayaki* there in the 1960s was not to Doppo's tastes. But, it would seem, Junichirō Tanizaki had a quite different impression when he said that Tokyo people should try "Hyōtei's cuisine" at least once.

In response to Doppo, locally born and raised gourmand Jinnosuke Inoue, who had his own views on cooking and restaurants, acknowledged that "Hyōtei, Tsuruya, and Chimoto are first-class restaurants in Kyoto," but "when one is talking about eel, one cannot leave out the *ochazuke-unagi* at Kanemasa in Sanjō Nawate."[45] He devotes a considerable amount of time to writing about *kabayaki*:

> One can always rest assured when eating at Okazaki Tsuruya, no matter when one visits. The balance of seasoning is appropriate, although perhaps somewhat on the heavy side, every dish will be enjoyable to some degree. And I always marvel at how surprisingly good the *kabayaki* eel served at the end of the meal is. Even hearing mention of eel *kabayaki*, let alone seeing it, after eating and drinking to such a degree, is enough to put one off, but the glutton that I am, I always look forward to it and devour the whole thing. That's just how difficult the taste of the eel here is to overlook.[46]

Inoue recalls that as a child, "I couldn't wait for the thick *kabayaki* eel that always came out at the end" at Heihachi Jaya in Yamabana.[47] This and the previous description suggest it may have been customary for restaurants in Kyoto to serve *kabayaki* as the final course of a meal.

Inoue enjoyed the Edo-yaki of Kyoto, "although its flavor may be far removed from the real taste of eel from Komatsu in Tokyo, I do enjoy the eel from Umenoi in Kyoto in my own way."[48] The author Kaisen Iguchi similarly recalls that "the flavor of Komatsu left an impression on me, in Tokyo, the home of *kabayaki*."

> Rosanjin Kitaōji, in his book *A Great Culinary Nation*, states that "there are two methods of grilling eel: the provincial method of direct grilling and the steam grilling used in Tokyo. And there are no two ways about it, the steam grilling of Tokyo is better." However, I feel that directly grilling wild eel is better, perhaps because I am a child of Kyoto.
>
> This year, it will soon be the Day of the Ox [i.e., Midsummer's Day].
>
> I would really like to go to Heihachi Jaya in Yase and eat wild eel that has been grilled directly.[49]

Iguchi is expressing his fondness for the Kansai-style *kabayaki* served at freshwater fish restaurants.

Everyone's tastes are different. In writing this book, we the authors also discovered that our tastes in *kabayaki* differ markedly from each other.

Notes

1. Usui, Kinosuke, *Kyoto mikaku sanpo* (A gustatory stroll through Kyoto), Shirakawa Shoin, 1963 (5th edition), p. 171.
2. Kinosuke Usui mainly discusses restaurants in *Shinpen, Kyoto mikaku sanpo* (A gustatory stroll through Kyoto: New edition) (Shirakawa Shoin, 1970), mentioning that "A well regarded *Kyō-yaki* eel restaurant is Kamada in the Shinagawa Food Center on the south side of the western approach to Nakadachiuri Shichihonmatsu." p. 483.
3. Hayashi, Nozomu, "(Sakka no kofuku) Karikari unagi" ((Fortunate flavors) Crunchy eel) (*Asahi Shimbun*, January 5, 2019).
4. Japan Society for Historical Research of Manners and Customs, ed., *Zusetsu: Edo jidai shoku seikatsu jiten (shinsō ban)* (The illustrated encyclopedia of dietary life in the Edo period (New edition)), Yūzankaku Shuppan, 1996, p. 32.
5. Usui Kinosuke, *Kyoto mikaku sanpo*, Shirakawa Shoin, 1962 (1st edition), p. 164.
6. Kansai Scenic Spots and Historic Sites Survey Committee (Komaki Shōichi) ed., *(Mottomo wakari yasui) dai Kyoto shi kankō annai* ((The easiest to understand) Sightseeing guide to the great city of Kyoto) (Kansai Scenic Spots and Historic Sites Survey Committee, 1932). Quote from Appendix "Dai Kyoto meibutsu meisan chomei shōten shōkai enkaku shi" (Introductory history of famous business establishments with Kyoto specialties).

7. Fujisawa, Takeo and Kitano, Eizō, *Kyoto, Osaka, Kobe: 300 en aji no mise* (Establishments of depth: Kyoto, Osaka, Kobe), Yuki Shobō, 1966, pp. 120–121.
8. Matsukawa, Jirō, *Zenkoku kagai meguri* (Nationwide tour of entertainment districts), Seibundō, 1929, p. 120.
9. Isobe, Shizuo, *Nezu yūkaku shō ki* (An account of the Nezu pleasure district), Edomachi Mei-rizoku Kenkyūkai (reprint, Kasutori Shuppan, 2017), pp. 11–12.
10. *Yomiuri Shimbun*, August 10, 1887. However, it was later reported that "there are plans to make it the number one hot spring in Tokyo, not only through licensed prostitution but also as a place for beautiful women to courteously entertain guests from home and abroad" (*Yomiuri Shimbun*, November 2, 1887).
11. *Yomiuri Shimbun*, July 8, 1884.
12. *Yomiuri Shimbun*, September 25, 1884, September 16, 1886.
13. *Tokyo Asahi Shimbun*, October 3, 1888.
14. *Yomiuri Shimbun*, *Tokyo Asahi Shimbun*, October 6, 1888.
15. Horiguchi, Seiichi ed., *Jōshū Isobe onsen annai* (Guide to the hot springs of Jōshū Isobe), Horiguchi Seiichi, 1889, pp. 13–14.
16. *Yomiuri Shimbun*, July 6, 1889.
17. *Tokyo Asahi Shimbun*, August 18, 1889.
18. Kato, Masahiro, *Hanamachi: Ikūkan no toshi shi* (Entertainment districts: A history of an otherworldly urban space), Asahi Sensho, 2005.
19. *Yomiuri Shimbun*, April 4, 1891.
20. *Tokyo Asahi Shimbun*, September 21, 1890.
21. See the following *Kan'inroku*: Nishimura Juntarō, ed., *Kan'inroku: Meiji 9 nen 9 gatsu* (Kan'inroku: September 1876), Nishimuragumi Shuppankyoku; Higurashi Tadatsune, ed., *Kan'inroku: Meiji 10 nen 12 gatsu* (Kan'inroku: December, 1877), Enryusha; Ōsaki Zenshirō, *Meiji Kan'inroku: Meiji 11 nen 8 gatsu* (Meiji Kan'inroku: August 1878); Yajima, Hyakutaro and Hikone, Shōzō, eds., *Kaisei Kan'inroku: Meiji 12 nen 1 gatsu* (Revised Kan'inroku: January 1879), Hakkō Shoin, etc.
22. *Yomiuri Shimbun*, November 18, 1894.
23. Tsujimoto, Jisaburō, ed., *Kyoto annai miyako hyaku shu (zōho ni ban)* (Guide to the one hundred faces of Kyoto (Enlarged second edition)), Shōtokukan, 1894, p. 208.
24. Kokubu, Ayako, "Gion Umenoi" (Sōgensha Editorial Department, ed., *Kyoto mikaku chizu: 1975 nen ban* (Kyoto taste map 1975 edition), Sōgensha, 1975) pp. 64–65.
25. Irie, Mikizō, *Unagitsū*, Shiroku Shoin, 1930, pp. 139–149.
26. *Tokyo Asahi Shimbun*, June 5, 1896.
27. Gion Umenoi website (https://umenoi.jp/), last viewed March 14, 2020. Another guidebook by Ayako Kokubu (*Zoku, Kamera: Kyō aji 100 sen* (100 selections of Kyoto flavors: Camera edition), Tankō Shinsha, 1964, pp. 62–63) and Kinosuke Usui's *Kyoto mikaku sanpo* (1st edition, pp. 168–169) mentioned above, both also calculate the year as 1919.
28. Usui, *Kyoto mikaku sanpo* (1st edition), pp. 168–169.
29. Dōmoto, Kansei, "Meiji no koro no Kyoto ryōri ya" (Kyoto restaurants in the Meiji era), (*Rakumi* vol. 100, 1960), pp. 34–36.

30. Usui, *Kyoto mikaku sanpo* (1st edition), pp. 166–167.
31. Kyo Meibutsu Hyakumikai website (http://www.kyomeibutuhyakumikai.jp/shop/kanemasa.html), last accessed April 11, 2020.
32. Usui, *Kyoto mikaku sanpo* (1st edition), pp. 166–167.
33. Irie, *Unagitsū*, p. 151.
34. Usui, *Kyoto mikaku sanpo* (5th edition), p. 171.
35. Kyoto Central Telephone Bureau, ed., *Kyoto shi denwa bangō bo* (Kyoto city telephone number book), Kyoto Central Telephone Bureau, 1938.
36. The abovementioned *Shinpen, Kyoto mikaku sanpo* also states that "Izumoya in Shinkyōgoku Nishiki retains its original style and serves Kyoto-style kabayaki," p. 213. However, Izumoya no longer exists.
37. Inoue, Jinnosuke, "Matsuno (unagi)" (Matsuno (Eel)) (Sōgensha Editorial Department, ed., *Kyoto mikaku chizu: 1969 nen ban* (Kyoto taste map 1969 edition), Sōgensha, 1968), pp. 58–59. Quote from p. 58.
38. This account of events was compiled based on: Usui, *Kyoto mikaku sanpo* (1st edition), pp. 172–173; "Sōhonke Nishin Soba Matsuba" website "About Matsuba / History" (http://www.sobamatsuba.co.jp/about/index.html), last viewed on April 14, 2020.
39. Kokubu, Ayako, "Matsuno" Sōgensha Editorial Department, ed., *Kyoto mikaku chizu: 1975 nen ban*, Sōgensha, 1975), pp. 88–89.
40. Matsuno website (https://unagi-matsuno.com/about/), last viewed August 16, 2021.
41. Inoue, "Matsuno (unagi)," p. 59.
42. Usui, *Kyoto mikaku sanpo* (1st edition), p. 170.
43. Kunikida, Doppo, "Saikyō ryōri shirōto hyō" (Amateur reviews of cuisine in western Kyoto) (*Kunikida Doppo zenshū: dai 9 kan* (The complete works of Doppo Kunikida, vol. 9), Gakkensha, 1966), pp. 690–695. Quote from p. 690.
44. Kunikida, "Saikyō ryōri shirōto hyō," p. 691.
45. Inoue, Jinnosuke, "Rakumi hanjōki – Kyō no tabemono, sono 2" (*Rakumi* vol. 70, 1957), pp. 62–165. Quote from p. 65.
46. Inoue, Jinnosuke, "Rakumi hanjōki – Kyō no tabemono, sono 10" (Rakumi Prosperity Chronicle: Food in Kyoto, Part 10) (*Rakumi* vol. 78, 1958), pp. 102–105. Quote from p. 102.
47. Inoue, "Rakumi hanjōki, sono 2," p. 64.
48. Inoue, p. 65.
49. Iguchi, Kaisen, "Unagi" (Eel) (*Tankō*, vol. 14, no. 7, 1960), p. 85.

10.2478/9784991351600-011

Chapter 4
Modern Chicken Hotpot

4.1 Junichirō Tanizaki and the Chicken Cuisine of Hanamikōji

4.1.1 Like Irikin of Mukōjima

> On an evening of fitful spring rain, having been invited by Mr. Ueda, I caught a rickshaw with Nagata to Hyōtei in the precincts of Nanzenji Temple ...
>
> Led by a young maid dressed in simple cotton *futako* [twice folded yarn] padded garments, I leaned against the eaves of the main building as the raindrops trickled down it. As I walked around to the rear garden, *and from the house's construction in the style of "Irikin," which is typical of teahouses in Kyoto,* I presently came to understand that this was Hyōtei.[1]

In April 1912, Junichirō Tanizaki and Mikihiko Osada were invited to Hyōtei by Bin Ueda, where they drank excellent sake to the point of intoxication, as mentioned in Chapter 2. The rickshaw they hired pulls up "in front of the eaves of a shabby house that seemed like an *imobō* restaurant," so Tanizaki assumes that "presumably, the rickshaw driver is going to buy candles or straw sandals there." Thus, he is somewhat puzzled to learn that this is the front entrance of Hyōtei but is nonetheless satisfied with the scene when he goes inside the building.

He comments that Hyōtei was built "in the style of 'Irikin,'" a famous restaurant in Mukōjima, Tokyo, that was "known among sophisticates and discerning customers in the Meiji era."[2]

> In the midst of green pines and verdant bamboo of various heights, this modest pavilion is inordinately well appointed, suitable for nobility and elegant guests to engage in flower viewings at dawn in spring, moon viewings of an autumn evening, and other excursions. It offers both cuisine and accommodation, with more than 20 guest rooms, both large and small, which are interspersed between springs, rocks, plants, and trees. Quiet, elegant, and refined, it is a truly singular paradise in Kyoto.[3]

Irikin was founded around 1908 by Kaneko Iriyama (whose nickname was "Kin"), who was well-known among women proprietors of the era. Her name is often mentioned alongside Kin Itō of the restaurant Shinkiraku in Tsukiji 3-chome and Kin Okada of the restaurant Okada in Hamachō.[4]

Tanizaki, an up-and-coming novelist in his early 20s, barely past his literary debut, must have visited Irikin in Mukōjima. It was his first time in Kyoto, so it was perhaps inevitable that he would compare the construction of the restaurant with those in Tokyo. Although Tanizaki describes the style as "typical of teahouses in Kyoto," the teahouses in both Pontochō and Gion are packed in closely together in the style of traditional Kyoto townhouses, so guest rooms in typical teahouses in Kyoto were unlikely to have been "interspersed between springs, rocks, plants, and trees." In fact, he visited a restaurant, Kikusui, where he experienced "the Irikin style" before Hyōtei, thus he describes the style is "typical of teahouses in Kyoto." Kikusui was located on the south side of Gion (on Hanamikōji), where he was invited by a local Kyoto sophisticate to eat *tori-nabe*.

4.1.2 What is *tori-nabe*?

In his *Suzaku Diary*, Tanizaki wrote: "accompanied by Kaneko – the young proprietress of the ryokan Yorozuya on Sanjō-dōri – in the evening, I went to eat chicken cuisine at a restaurant called Kikusui on Hanamikōji." In later years, he recalled the scene as follows:

> Nowadays, the neighborhood in which Kikusui is situated is full of teahouses and geisha houses, but back then, I remember there was just a single restaurant, Toriya, standing alone in a quiet area like it was in the middle of an empty field. It may have been just that time when Hanamikōji was about to become a busier place. I remember that Shigenao Chaya was still in the northern part of Shijō in the direction of Shinbashi, but Mantei was in the same place as it is now, at the corner of Hanamikōji.[5]

This passage suggests that the area on the southern side of Gion was still under development at the time. In spring 1912, Shijō-dōri underwent "expansion works" to simultaneously widen the street and install streetcars. During this construction, "seki-zashiki" ("rental rooms," i.e., teahouses) facing Shijō-dōri and Nawate-dōri were not allowed to operate,[6] so many teahouses relocated to the southern side of Gion. A succession of new establishments opened there throughout the Taishō period (1912–1926), creating the landscape of the entertainment district that persists to this day. The chicken cuisine restaurant Kikusui was something of a pioneer, at least in terms of restaurants in this southern area of Gion.

> Tanizaki visited Kikusui with the poet Mikihiko Nagata and others during this period:
>
> Speaking of Kikusui, I am reminded of Toriya in Ningyōchō, but the structure of the house is quite different. Following the stepping stones and parting the thick shrubbery,

> I arrive at the inner garden. Here, on the shores of a modest pond, two or three small tatami rooms reminiscent of Irikin in Mukōjima can be seen amid the twilit green leaves.

Here, Tanizaki once again mentions Toriya. He may be referring to the Kikusui (Kyoto fish and chicken dishes) located in the (former) Shin-Izumichō, which is now part of Ningyocho 3-chome.[7] The sight of the separate rooms like those of a tea room reminds him of Irikin in Mukōjima.

With a guide well-acquainted with Kyoto's *kagai* entertainment districts, there were plans for the restaurant to go well beyond "chicken cuisine" this evening.

> It was an uncomfortably humid and hot night, with a thin layer of clouds. Each taking turns to fan ourselves, we all gathered around the pot and drank from our cups. The sake and chicken were both quite good.
>
> "Thank you for tonight..."
>
> A maiko from Gion greeted us charmingly in the Kyoto dialect as she entered the room, her obi tied up like a cat's toy. Two geishas followed her into the room.
>
> "What is this?" the maiko asked with a quizzical air, as she picked up the metal fittings used to remove the iron pot from the stove.
>
> "Ah I see. If so, then shall I do it? Please, allow me."
>
> Whereupon she removed the pot from the stove and then placed it back on again several times. When I enquired as to her age, she replied:
>
> "I'm 13."
>
> She quietly lowered her head and fiddled with her dance fan.[8]

As we can see, this was a banquet of sufficient largesse for geisha and maiko [an apprentice geisha; lit. "dancing child"] to attend. Tanizaki also recounts that the maiko's actions spurred a lively discussion on the different cultures of *kagai* in the eastern and western parts of the city. Subsequently, he moved across the river to Shinbashi, on the north side of Shijō-dōri, where he had a yet deeper tatami room experience. However, let us return to the scene above, focusing on the presentation of chicken cuisine.

Tanizaki describes sitting in a small room in Kikusui, which reminds him of Irikin; he is seated around a "pot" and enjoying chicken cuisine, with which he seems satisfied: "The sake and chicken were both quite good." The exchange with the maiko also clarifies that the pot in question was an "iron pot."

That being so, what kind of dish exactly was this "tori-nabe" that they were grazing over while quaffing drinks?

> Torishin in Kyoto's Sanjōji-chō is a good place for *tori-nabe*. There have been rumors that Torishin uses rabbit meat, but that is only because it is doing so well that people are jealous of it. Being so cheap and so delicious, it is only natural that the place would thrive.[9]

In the early 1920s, Tanizaki mentions this specific restaurant as an example of "Food in the Kyoto Area," once again discussing "tori-nabe." Aside from winning over the harsh critic Tanizaki, though, the specifics of this *nabe* hotpot dish remain unclear, aside from the fact that chicken is the main ingredient. To learn more, we will need to go back to the early Meiji period and examine restaurant menus.

4.2 "Wa" or "hiru"?

4.2.1 Signboard menus

Pioneers of the Capital, a guide to commerce and industry published in 1883, featured some of the leading restaurants in Kyoto at the time in its "Culinary Business" section.[10] The entry for "Uotō, northwestern end of Sanjō Ōhashi Bridge" (Fig. 4.1) includes an illustration of a restaurant with a signboard that reads "saltwater and freshwater fish," opposite another restaurant in a towering three-story building that is drawn in an exaggerated manner. Characters that read "hiru kashiwa" can be seen on the lantern that hangs outside the restaurant. In the upper left corner of the drawing, "kashiwa" is written. The character on the lantern that looks like a phonetic "wa" (わ) in "hiru kashiwa" is the character meaning "king" (王) written in a highly stylized form.

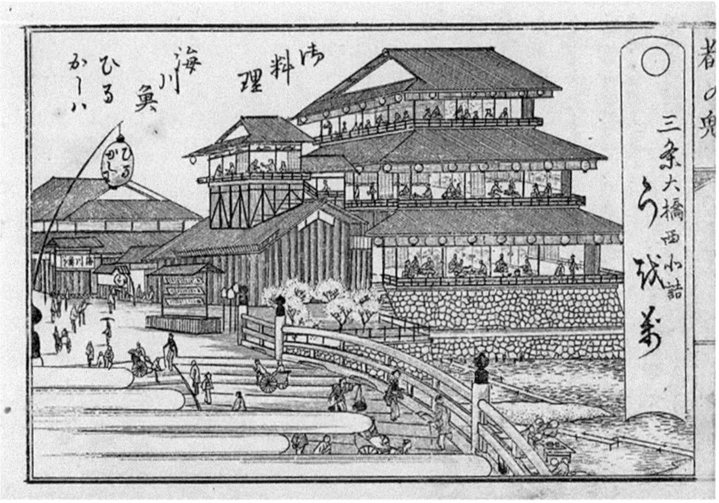

Figure 4.1 Uotō, at the northwest end of Sanjō Ōhashi Bridge

This combination of "kashiwa hiru" (same as hiru kashiwa) appears frequently in the culinary business section of *Pioneers of the Capital*. The entry for Mishimaya in "southern Senbon-dōri Imadegawa" also contains the phrase "kashiwa hiru" under "kaiseki" alongside "saltwater and freshwater cuisine." "Kashiwa" is a term still used today, meaning "chicken meat," but what does "hiru" mean? Forestalling the answer for a moment, let us look at the illustration of Toritoyo's entry in "northern Shinkyōgoku Nishiki" (Fig. 4.2).

Take a look at the large list of menu items written in the top left of the image. Reading from right to left and top to bottom, the items are as follows:

> Hiru, Kashiwa, Unagi [eel], Manbushi, Chawanmushi, Maki-tamago, Umaki [a kind of eel sushi], Dojō [loach], Koi-funa [carp], Osake, Gozen [traditional formal table service]
>
> "Manbushi" means "eel rice" in the Osaka–Kyoto area. Then, there are various egg dishes, such as *chawanmushi,* which is a savory egg custard, and *maki-tamago*, a seasoned omelet that has been rolled up.

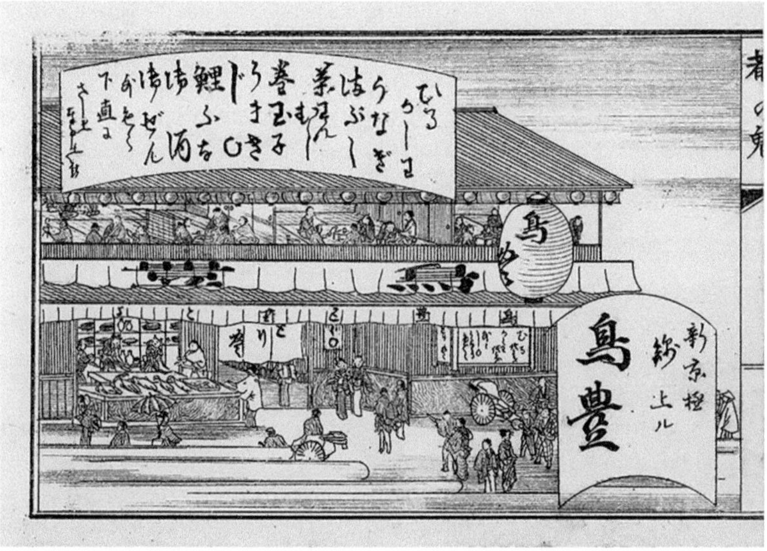

Figure 4.2 Toritoyo, northern Shinkyōgoku, Nishiki

However, "hiru" and "kashiwa" once again appear at the top of the list. Whatever "hiru" may mean, it seems that serving chicken and freshwater fish dishes was a characteristic of restaurants in Meiji Kyoto.

4.2.2 The "laziness boxes" of kashiwaya

In Kyoto in the Meiji era, "kashiwaya" was established as a genre of restaurant similar to "shamoya" in Tokyo, "shamo" being another term for chicken. Centered around the *kagai* entertainment districts of Gion and Pontochō, the *kashiwaya* were so widespread that "there are such places everywhere."

A guidebook on nightlife in Meiji Kyoto provides an interesting discussion of "kashiwaya" which is worth examining at some length.[11]

> There are such places everywhere, mainly in the area around Gion and Pontochō, that are equivalent to what is called "shamoya" in Tokyo. So, let us here provide a rough illustration of one of the best among them. First, after passing through a half-*noren* curtain, a waitress bowed deeply, touching her forehead against the threshold of the building, and guided us to a room with plastered walls. We passed through much of the room, and all I could see in the space, which was partially hidden in the light and shadows cast by the electric lighting, were a large vase that was perhaps not quite in style, with a budding camellia or something inserted therein, a dubious scroll with a somewhat faded old-fashioned look, what was perhaps an incense burner of unclear provenance, and a cluttered array of writing implements. This was the appearance of the room at a glance.

This description of the furnishings of the restaurant conveys a clear sense that the restaurant has been converted from a townhouse. The guest who has been shown into the room is forced to make a choice.

> After we had settled down on the cushions proffered by the waitress, having decided on who was sitting where, the waitress asked us ... "Would you like 'wa'? Or would you like 'ru'?'" In response to which, we were somewhat bewildered; although we were not "o-agari-san" (the general term for those who come from various places to see the sights of Kyoto), we were still not accustomed with these terms. We asked her what on earth she might mean, whereupon she informed us that "wa" is an abbreviation of "kashi**wa**" and "ru" is a synonym for "ahi**ru**" [duck] used in Kyoto.

Here, at last, we get a glimpse of the meaning of "kashiwa hiru." Sources note that "until the Meiji period, the *kashiwaya* of Osaka had lanterns bearing the words 'hiru kashiwa'"[12] or that "at poultry restaurants in the Kansai region, 'hiru, kashiwa' was written on their signboards and lanterns, with 'hiru' being an abbreviation for *ahiru*."[13] The discussion above indicates that the phrase "kashiwa hiru" was used outside Osaka and more recently than the Meiji period. Meanwhile, "ahiru" refers to a cross between domestic and wild ducks. The Kikusui on Hanamikōji, which Junichirō Tanizaki had found so delicious in the Meiji era, also advertised itself as offering "chicken and duck dishes" in the 1920s.[14]

After the customers had finished ordering, the waitress brought them a box.

> ... all the dishes that would be served on a large tray in Tokyo, from grilled tofu, *konnyaku*, green onions, dropwort or *mitsuba*, grilled wheat gluten, dashi stock, soy sauce, mirin, to sugar, are all placed within a single piece of lacquerware like a tea cabinet, which is all

> of a size where one need only open a drawer to bring out the chopsticks. This is called a *bushōbako* ["laziness box"], probably because one can do everything while sitting.

Thus, at *kashiwaya* in Kyoto, a wooden box called *bushōbako* containing all the ingredients and seasonings was brought to the seated guests (Fig. 4.3). These ingredients (grilled tofu, *konnyaku*, green onions, water dropwort, and fried gluten) and seasonings (dashi, soy sauce, and sugar) suggest that "chicken cuisine" or "tori-nabe" referred to *sukiyaki* – a dish comprised of thinly sliced meats and vegetables cooked in soy sauce and sugar. It appears that Tanizaki was grazing on chicken *sukiyaki* at Kikusui.

Figure 4.3 Picture and photograph of a *bushōbako*

However, in present-day Kyoto, it is rare to find *sukiyaki* at restaurants specializing in chicken dishes; many restaurants have changed the menus shown on their signboards from *sukiyaki* to *mizudaki* (chicken hotpot). It is no exaggeration to say that chicken *sukiyaki* is now rather uncommon in Kyoto cuisine. But as we can see in the following quote from kabuki actor Kataoka Nizaemon XIII (1903–1994), both dishes were widely available in postwar Kyoto:

> Chicken *mizudaki* is also delicious, at Shin Miura in Gion and Kiyamachi Nijō, or the chicken *sukiyaki* at Toriyasa in Kiyamachi.[15]

Toriyasa, founded in 1788, is located on the right bank of the Kamo River (southern Kiyamachi Dankuri Bridge). It was so famous that it would have been the first place suggested to anyone looking for chicken *sukiyaki* at the time when Kataoka visited. And today, Toriyasa specializes in *mizudaki*.

Kataoka also mentions Shin Miura in Gion Shinchi and Kami-kiyamachi, praising their chicken *mizudaki* as "delicious." This Shin Miura is in fact the restaurant that first brought *mizudaki* to Kyoto. To understand how that came about, let us make a flying visit to the city of Hakata in search of the roots of *mizudaki*.

4.3 Former brothels and chicken *mizudaki*

4.3.1 The gastronomy of Hakata

> Hakata is not only a center of art, but also a city of gastronomy. There are signboards for small restaurants, *mizutaki*, and *fugu* [pufferfish] dishes everywhere. Sushi restaurants never fail to get bottoms on seats at their counters.[16]

Hakata – the "city of gastronomy." One of the dishes that gives Hakata this reputation is *mizutaki* (this dish is variously referred to as "*mizudaki*" and "*mizutaki*" using different characters in Japanese, which will be explained later). About 60 years ago, the guidebook *Journeys in Hakata: Night and Day* proclaimed that "Nagoya to the east and Hakata to the west are the *yokozuna* [top-ranked sumo wrestlers] of east and west chicken cuisine." In the Postscript, the guide's editor comments:

> Although the city has beautiful spots for sightseeing, it also often shows its grubbier side. It would be a disservice to travelers to introduce them to the famous and historical sites of my hometown Hakata, without also telling them about the beautiful evening lights and the reality of the murky back streets.

As we can see, the guidebook includes both the hard and soft aspects of the city, detailing a nightlife that is not always pretty, and "the reality of the murky backstreets." Although we may be tempted to wander the back streets as we follow this guide, let's stick on course with *mizutaki*, which is part of the prettier side of this book, the "Eating Tour: Delights for the Tongue" section.

> No matter where you go in winter in Hakata, you will be assailed with *mizutaki* on all sides as a matter of course. Standing in front of small restaurants here and there are signboards that read *mizutaki*, ranging from those such as Ippaiya, catering to humble folk offering a one-pot meal 200 yen per person, to high-class eateries where the waitress will cook the meal [before your eyes] while geisha pour your drinks.

Prices are not what they were 60 years ago, and with the decline of the *kagai* geisha districts, the idea of a geisha pouring your drinks seems almost unimaginable. Although the word *mizutaki* can still be seen on the signboards of bars, "*motsunabe*" (beef or pork tripe hotpot) is much more commonly seen in Hakata today. Originally, *mizutaki* season was considered to "begin in fall and end in spring," but as air-conditioning became more widespread in restaurants, it became a year-round dish, and "*mizutaki* in the heat of summer" became "a fine thing."

Another guidebook observes:

> When it comes to the flavor of Hakata, *mizutaki* must be mentioned first. Walking around the city, you will see many signboards offering *mizutaki*, with prices ranging from 300 yen per person to over 1,000 yen.[17]

This foodscape and streetscape is what makes Hakata a "city of gastronomy."

Among the restaurants offering *mizutaki* at various prices, Shin Miura is the first to be mentioned: "*Shin Miura in the old Yanagimachi district* is the most famous, with its specialty of *mizutaki*," described elsewhere as "first among the famous chicken specialty restaurants of Hakata."[18] According to the comedian Roppa Furukawa:

> Hakata. For lunch I had the *mizutaki* of Shin Miura, which I would *rather call a famous landmark* than a specialty; eating there made my pulse pound and made me want to let out a moan of delight. I chewed on the bones with such furious abandon that my mouth bled.
>
> Shin Miura is probably the best in the world for *mizutaki*.[19]

This quote is from an article that originally featured in *Umeda Goraku Shimbun* in July 1940, so can be considered as an impression from the early Shōwa period.

Shin Miura in Hakata is a long-established restaurant located at the base of Hakata Port, on land that has now been reclaimed, on the left bank of the mouth of the Mikasa River (Ishido River), which flows through the northern part of the old city. Its premises were rebuilt after the war and then demolished in 2006 and replaced with a new one-story building occupying about half of the original site. Surrounded by buildings and condominiums and with an urban highway running across the opposite bank, the location can hardly be called scenic. There are no noteworthy restaurants in the surrounding area, and it undeniably feels like an isolated place.

However, before the war, Roppa had good reason to praise it as a "famous landmark."

4.3.2 Old Yanagimachi and Shin Miura

Two guidebooks describe Shin Miura as a famous restaurant, but one locates it in the Chikkō port area and the other in "the old Yanagimachi district." Of course, if there is old, there is also new.

> There is a folk song, long known across the country, that goes "there are no willows [*yanagi* in Japanese] in Yanagimachi, Hakata, only wenches with willow waists." Speaking of this, those who come from other regions to frequent the pleasure districts count these wenches one of Hakata's specialties. Hakata Port has long been a port of

> commercial importance in the Saikai Sea, where many ships come and go, and its system of courtesans was established at an early date. During the Tenshō era [1573–1592], the area around what is presently Suzaki-chō was already lined with brothels, and the young wenches of Hakata were already known throughout [the Saikai Sea]. Then, during the Keichō era [1596–1615], the brothels moved to Ishidō Kawajiri, which is the current Yanagimachi. In around Meiji 24 [1891], the beach was reclaimed to expand the area, showing the extent to which the nightlife district was thriving. However, in Meiji 43 [1910], it was once again moved to a spacious district in Sumiyoshi outside the city and referred to as the brothel district of *Shin-yanagimachi* [New Yanagimachi].[20]

Thus, Yanagimachi was the historic red-light district. The brothel functions were relocated to newly developed land on the left bank of the Naka River, due south from the mouth of the Ishido River across the central districts of Hakata, with the former site becoming "Old Yanagimachi" and the new site becoming "New Yanagimachi."

Underlying this need to relocate Yanagimachi, as it were, antipodally, from the northern end of the city to (what was at the time) the southern outskirts was the establishment of a medical college (now Kyushu University School of Medicine).

> Around Meiji 40 [1907], little by little the issue of moving Yanagimachi began to arise. With the different prefectures of Kyushu vying with each other to be the site of the medical school, Fukuoka Prefecture won with a decisive victory. It was decided that Chiyo-matsubara would be a good location for the site, with its white sands and green pine trees. This site, however, was only separated by a river across from the gay district of dark willows and bright flowers, which was quite a headache for the authorities, who were concerned that it would be unfavorable in terms of public morals for students tasked with exploring the origins of life and the innermost workings of disease to be thus situated and therefore insisted on the relocation of Yanagimachi at all costs.[21]

A medical college on the Ishidō River, located "across from the gay district of dark willows and bright flowers;" after Nezu in Chapter 3, this is the second time we've mentioned brothels being forced to relocate due to the emergence of modern institutions of higher education. Will there be a third time? You may well wonder...

A topographical map from the mid-Meiji period depicts the mouth of the Ishidō River (Mikasa River) where Yanagimachi was located on the outskirts of the city at that time.[22] Sandbanks of accumulated sediment have formed near the mouth of the river, with sandy beaches and pine forests all the way to Najima on the opposite bank.

By the early to mid-1920s, streetcars were running across a new bridge over the Ishidō River and the Hakata Bay Railway had been built along the northern beach, with Shinhakata station at the junction between the two. The medical college occupies a large site to the northeast of the station. Although there is a place named Chiyo-matsubara (lit. "Plain of long-lived pine trees") on the east side of the old Daigaku-dōri (University Street), with symbols on the map indicating coniferous forests (pine trees) scattered around the area of Higashi Park, the progress of development is remarkable, hinting at further expansion of the urban area. Following the decision to establish the medical college, the brothels (*sekigashi* rental rooms) of Yanagimachi were moved to Sumiyoshi, which was outside the city at that time. The new brothel district opened swiftly in June 1909.[23] This was the birth of Shin-yanagimachi (Fig. 4.4).

Most of the *sekigashi* establishments moved to Shin-yanagimachi. Old Yanagimachi literally became "a district where the lights have gone out," except one establishment kept the lights on: Shin Miura, was alone in the area and changed its business from *sekigashi* rental rooms to a restaurant. According to Jirō Matsukawa:

> Shin Miura is a *mizudaki* establishment, but *it is the only house in old Yanagimachi that remains of the old brothels.* Swaddled by the Ishidō River and overlooking Matsubara on Umi-no-Makamichi in the distance, it is eminently suitable for a light drink on a moonlit night, and the chicken is also the pride of the establishment.[24]

Figure 4.4 View of Shin-yanagimachi

Jirō Matsukawa was a travel writer with a thorough knowledge of *kagai* entertainment districts throughout Japan. This description of the location and view may have been what prompted Roppa to describe Shin Miura as a "famous landmark." The building, having been converted from a *sekigashi* establishment, was quite splendid, with a corridor running through the central garden and a large hall in the main building (Figure 4.5).

Matsukawa declared that "chicken *mizutaki* is the only specialty, and the Genkai sea bream also has a reputation for being delicious,"[25] giving some clue as to what the foodscape of Hakata and her twin city Fukuoka was like when Shin Miura opened in 1910. Another clue can be found in the *Fukuoka City Guidebook* (1910), which describes the phenomenon known as "kuidaore" (literally, "food downfall") at the end of the Meiji era:

Figure 4.5 Exterior and inner garden of Shin Miura

> We say in Chikuzen [the former province that included Hakata/Fukuoka] that one can bring ruin upon oneself through extravagance in food. After all, there are people of Chikuzen, especially people of Fukuoka, who have devoted too much attention to selection of foods and drinks that they lavishly spend tens of yen on rare delicacies for the dining table, leading ultimately to the loss of their lives and the downfall of their families.
>
> However, admittedly the food here is made that much more delectable by time-honored cooking methods, the unrivaled flavor of seafood of the Genkai Ocean, with its firm flesh, and well-grown vegetables aided by fertile soil.[26]

Naturally, Shin Miura is not yet listed among the restaurants therein, and there is no mention of *mizutaki*. It seems that Matsukawa first learned about *mizutaki* during the subsequent Taishō period.

> I think I'll go to eat *mizutaki*. Nowadays, one can also get *mizutaki* in Tokyo, but it's nothing like eating it in Hakata. They say that they bring in chefs from Hakata, but the customers don't know what *mizutaki* is, so it seems that they make some allowances to suit the palates of Tokyo people. That's why we don't want people thinking that Tokyo's *mizutaki* is the real thing.[27]

When he spent a spell in Hakata on a "tour of nightlife districts," which likely also doubled up as a tour of *kagai* across Japan, on the recommendation of a friend who worked at a local newspaper company, Matsukawa learned about the "real *mizutaki*" of a restaurant called Suigetsu, which is credited as the birthplace of Hakata *mizutaki*. He likely also became acquainted with Shin Miura on this tour of the *kagai*. Einosuke

Obiya (1916–1993) claims that *mizutaki* originated from iriyaki," a hot pot dish found today in Nagasaki and that it was "Nanban cuisine", which means it was influenced by non-Japanese sources, most likely Portuguese or Chinese. He goes on to explain:

> It was Suigetsu (in present-day Hirao Honmachi, Fukuoka) that made it into the Japanese-style *mizutaki*, but it was Shin Miura that made it famous by opening a *mizutaki* restaurant in the middle of what was then a pleasure district.[28]

Furthermore, according to Einosuke Obiya, the originator of the dish, Suigetsu, started selling *mizutaki* in Dotemachi on the left bank at the mouth of the Naka River in 1905.[29] Obiya declares that "the *mizutaki* is delicious because the soup is good (there are two types: potage-style and consommé-style) and because the meat is good and falls off the bone easily," although it was Shin Miura in old Yanagimachi that "made *mizutaki* famous" with its cloudy "potage-style" soup. But, Obiya explains, Shin Miura

> did not do well at first, and apparently the only thing that sold was *karaage* fried chicken. However, it soon became popular after Kyushu Imperial University (now Kyushu University) opened nearby. This was because it was cheap, fashionable, and nutritious, so everyone from students to professors would flock there in droves every day.[30]

This raises many questions: Was fried chicken common on menus at that time? Was the chicken really that cheap? *Mizutaki* is a hotpot dish, so was it really ever "fashionable"? Nonetheless, it is true that Shin Miura "made *mizutaki* famous." The guidebook *Fukuoka City*, published in 1916 probably based on the *Fukuoka City Guidebook* (1910) quoted earlier, contains the following passage:

> Chikuzen has long had a reputation for "kuidaore," that is, bringing about one's downfall through extravagance in food. This, of course, insinuates that the people of Chikuzen, especially we people of Fukuoka, devote their hearts to food and drink and spare no expense on delicious meals, which can eventually result in misfortune for oneself and one's family. And it is in fact true that we developed our cooking methods early on, because the city of Fukuoka has fresh fish from the so-called Genkai Sea, whereas on land, there are fields of abundance, so there is never anything to be desired in terms of cooking ingredients.[31]

Although this passage on the "*kuidaore* of Chikuzen" is largely the same as the description in the *Fukuoka City Guidebook*, the following section in the newer guidebook has a new category of "chicken beef specialty and *sukiyaki*" with 'Shin Miura (Yanagimachi)' listed at the top. In 1914, the *Fukuoka Business Guide* featured an advertisement (Fig. 4.6), which corroborates the location.

Figure 4.6 An advertisement for Shin Miura

> Hakata Shokudōraku
>
> Where should we guide our customers to?
>
> First and foremost, (*kashiwa*) *mizutaki* is renowned across the land, with Shin Miura in old Yanagimachi being greatly praised [for this dish]. Moreover, Shin Miura's beautiful seating is a delight, where one can see everything from the room, an evening of spring rain, the color of the tide in summer, or the sea in late autumn.[32]

Thus, in the prewar part of the Shōwa period, the *mizutaki* of Hakata developed into a specialty dish that was advertised as being "renowned across the land," and Shin Miura in old Yanagimachi was the restaurant that made this specialty famous.

4.3.3 Onward to Kyoto

Shin Miura was founded in 1910 by Shūzaburō Aoyagi.[33] Chōkichi Aoyagi ran a *sekigashi* establishment called Miuraya in Shin-yanagimachi, and consequently, Shin Miura (literally "New Miura" derived from "Miura-ya") must have originally been a branch of this establishment. In the Taishō period, Shin Miura was also called "Shin Miura-ya."

The subsequent managers are listed in sources as Motoyoshi Aoyagi in 1914 and Asa Aoyagi in 1918.[34] As they all share the same surname, they were likely related to Shūzaburō in some way. Figure 4.7 presents two advertisements from the early Taishō era (the one on the left is from 1914 and the other is from 1916). The right-hand advertisement mentions a "Kyoto branch" in addition to the main restaurant in "Old Yanagimachi, Hakata," indicating that Shin Miura had advanced into Kyoto less than 10 years after its inception.

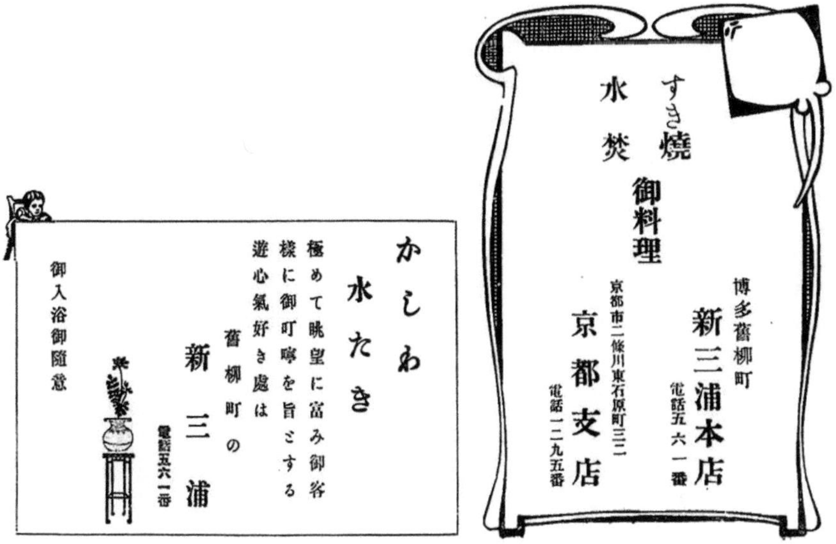

Figure 4.7 Taishō-era advertisements for Shin Miura (Left: 1914; Right: 1916)

There are two interesting stories about the establishment of Shin Miura in Kyoto:

> There is a theory that the original form of this chicken *mizudaki* was itself introduced from China to Hakata via Nagasaki. When I asked how the dish found its way to Gion Shin Miura, I was told that a woman named Asa Aoyagi, who had a connection with the original Hakata Shin Miura, brought her knowledge of how to make *mizudaki* and opened a restaurant in the eastern end of Marutamachi Kawabata.[35]

The above passage includes geographical information such as "Gion" and "eastern end of Marutamachi Kawabata," but what is more surprising is the way that Asa Aoyagi is mentioned. She had much more than "a connection with the original Hakata Shin Miura" and was in fact listed as the manager of the main restaurant in the Fukuoka business directories at the time.[36]

We will discuss the process by which Shin Miura was established in the next section. For now, let us see another account of its establishment:

> It seems that it was in Taishō 4 (1915) that Shin Miura advanced from the home of *mizutaki* in Hakata into Kyoto and opened its first shop in Ebisugawa Nijō.[37]

"Ebisugawa Nijō" appears to refer to a location on Nijō-dōri street near the Ebisugawa Power Station, and the date is clearly stated as 1915. This date is significant because it matches the advertisement in Figure 4.7. Thus, we can be sure that the Kyoto branch of Shin Miura was established in 1915.

As far as we know, the first appearance of the word "mizudaki" in Kyoto was in an advertisement for the restaurant Momijiya (Fig. 4.8), which was featured in a pamphlet for the Miyako Odori (an annual geisha performance) of 1915. The Miyako Odori is held every April, but in 1915, it was also scheduled to be performed in November to mark the ascension of Emperor Taishō. As was the case with the later ascension of Emperor Shōwa (Emperor Hirohito), Kyoto experienced a boom in consumer culture and advancements in the development of its urban space infrastructure. It seems that both Shin Miura and Momijiya opened their Kyoto branches to coincide with the enthronement. We note the similarity to when Kandagawa, with its Edo-style grilled eel, established a branch restaurant in anticipation of the Fourth National Industrial Exhibition (see Chapter 3).

Figure 4.8 Advertisement for Momijiya's "tori-mizudaki"

Let us now turn our attention to the location of the Shin Miura branch in Kyoto. What kind of location in Kyoto did this restaurant choose, with its origins in a former red-light district in Hakata?

The advertisement in Figure 4.7 has the address "32 Ishiwara-chō, Nijōgawa Higashi." "Nijogawa Higashi" was the name of a streetcar stop, and "Ishiwara-chō" was located a few minutes' walk east from there. However, there is no lot number 32 in Ishiwara-chō, so we cannot rule out the possibility that this is an error. The *Kyoto Telephone Directory* of 1919 has an entry for a branch named "Shin Miura Kyoto Shiten" under the address "4-chōme Ebisugawa Kawabata eastern entrance: Chicken mizudaki cuisine." The location may have been referred to as "Ebisugawa" because it was almost directly south of the Ebisugawa Power Station, located on the Lake Biwa Canal.

Looking up Ishiwara-chō on a map reveals that the location was near the former red-light district of Nijō Shinchi.

> Nijō Shinchi was licensed for prostitution along with Gion and has much the same origins as Gion, yet it was much less prosperous, both during the day and the nighttime, and many establishments moved to Pontochō after the area became licensed. Throughout the Meiji era, several schools were built in Yoshida-chō, and amid fears of harm to public morals if the schools were located near [to brothels], Governor Kitagaki decisively resolved to move [the brothels] to Kiyoi-chō in Shimogawara. However, they subsequently declined further, and now all such establishments in that district have been completely abolished.[38]
>
> Nijo Shinchi was originally a *kagai* in the 32nd School District of Kamigyō Ward and may have had connections with Chōmyō-ji Temple on the eastern end of Nijō Bridge or with Kumano Shrine. It was located between the current Ebisugawa Dam and the old railroad on Nijō... When Kyoto Imperial University was founded, the abolition of Nijō Shinchi was carried out with unexpected ease because of [the brothels' effects on] the educational environment.[39]

Article 1 of the Five Industry Control Regulations, promulgated in July 1886, states that Nijō Shinchi was a brothel district that had been certified as a "licensed area for *sekigashi* rental establishments" along with other historic entertainment districts. However, just two years later, on April 7, 1888, only "Nijō Shinchi" was abruptly deleted from Article 1 because of its proximity to the planned construction site for the Daisan Kōtō Gakkō (Third High School; later Kyoto Imperial University), as indicated in the above quote.

Nearly 30 years after Nijō Shinchi was abolished, Shin Miura of Hakata established a branch nearby, as though somehow retelling the tale of its own origins.

> Until the beginning of the war, when one passed through that area, there was something in the appearance of the area that was reminiscent of the days of Nijō Shinchi.
>
> For example, the main Kyoto branch of the *mizutaki* restaurant Shin Miura was located there. The *mizutaki* restaurant was not located there because it was near a dam but because it was a remnant of the Shinchi era. [Branches of Shin Miura] eventually appeared along the riverside in Kiyamachi but now seem to be found only in Gion Hanamikōji.[40]

The statement above diverges from the facts in two regards. One is that Shin Miura was "a remnant of the Shinchi era" and the other is that it was now found "only Gion Hanamikōji." We will discuss the second point in the next section, but the error in the first point is obvious: there is no evidence to suggest that Shin Miura of Hakata opened a branch during the "Shinchi era." However, the author of the above quote, Kōji Terao (1903–1995), who later became a professor emeritus at Kyoto

Sangyo University, must have had his own reasons for misapprehending that Shin Miura was a "remnant." The most likely reasons for his belief are the restaurant's location and appearance.

The former Nijō Shinchi area and its surroundings are shown in Figure 4.9. The areas shown in light gray indicate the extent of the existing urban area around 1887, when Nijō Shinchi was defined as Shin-ikesu-chō, Shin-pontochō, Daimonji-chō, Nakagawa-chō, Namba-chō, and Sugimoto-chō. This shows that Ishiwara-chō, where the Shin Miura in Kyoto was located, was on the edge of the existing urban area at the eastern end of Nijō-dōri and was located outside the Nijō Shinchi area. In the early Taisho period (1910s), houses and other structures were already being built in the area from the eastern side of Ishiwara-chō to the Lake Biwa Canal.

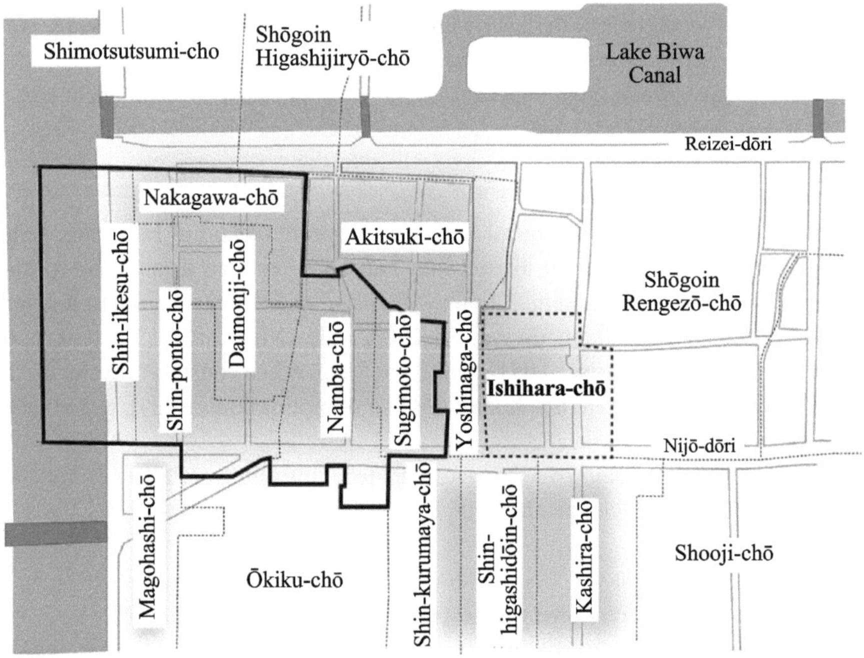

Figure 4.9 The area around Nijō Shinchi circa, 1887 (Meiji 20)

Note: The area enclosed with a thick black line is Nijō Shinchi.

Figure 4.10 shows two picture postcards depicting the Kyoto branch of Shin Miura. A signboard reading "Shin Miura Chicken Cuisine" hangs on the gatepost, and a lamp with "Shin Miura" written on it is above the main entrance on the arched gate. Two two-story wooden buildings can be seen within the high wall, which is protected by the traditional *inuyarai* bamboo fence at its base. The second image depicts a pond with a fountain arrayed with plants and ornamental garden stones.

Figure 4.10 Exterior and inner garden of Shin Miura Kyoto Shiten (Kyoto Branch)

Located at what was once the eastern limit of the urban area, Kōji Terao must have mistaken it for "a remnant of the Shinchi era" because it was such a grandiose building. Shin Miura was located close to the canal, which was admittedly an artificial body of water, and alongside a former red-light district. We can think of the selection of this location as arising from Asa Aoyagi's unique sense of place, derived from her involvement in managing the main restaurant in the old Yanagimachi district in Hakata.

4.4 The creation and spread of *mizudaki*

4.4.1 The development of Shin Miura Kyoto

The Shin Miura Kyoto Shiten branch restaurant became the main restaurant (Honten), among Shin Miura restaurants in Kyoto, in the latter half of the 1920s as two more branch restaurants were opened. Shin Miura first expanded into the southern side of Gion. Jinnosuke Inoue looks back on this time:

> I learned the taste of *mizutaki* long ago during my student days, and to this day, I cannot forget that it was in Shin Miura of Gion where I first tasted it. Shin Miura was located on the north side of the second street east of Shijo Hanamikōji. For some time after the war, there had been a temporary restaurant on Hanamikōji street, but when it finally returned to its original home, there was a sense of it having regained its composure. When it was on Hanamikōji, the location was so good that it ended up being difficult to get a table, but when it moved back to its original side street, it felt more intimate. To the east across the alley is what became Yasaka Junior High School, which used to be Yasaka Elementary School. I remember that one could glance down on the schoolyard from the window of the second-floor tatami room on the east side [of Shin Miura], and in summer, the sunlight would shine dazzlingly on the white, dry, dusty schoolyard. However, the building with second-floor tatami room was destroyed during the wartime evacuations, and the view of the schoolyard could be seen no more.[41]

This might suggest that *mizudaki* was eaten even at the height of summer in Kyoto. The building Inoue refers to first opened in 1925 on the south side of Gion, and was demolished in 2021. It was established by Asa Aoyagi as a branch restaurant named "Shin Miura Hanamikōji Shiten."[42] This is a clear indication of her preference for locations oriented toward *kagai* entertainment districts. Asa seems to have left the management of the main restaurant in Hakata around the time of the branch's opening, with the responsibility being passed on to Zenzō Shirai, Motoyoshi Aoyagi's brother-in-law.[43]

In 1927, the Shin Miura Hanamikōji Shiten was reorganized into the Shin Miura Gion Shiten, with Zenzō Shirai's younger brother Ryōzō as the manager, while Sōkichi Takahashi opened the Yase branch as mentioned in the introduction. Sources indicate that Takahashi was Asa's brother-in-law.[44] Thus, at that time, Shin Miura in Kyoto was managed by Asa Aoyagi in eastern Nijō Kawabata, Ryōzō Shirai in the southern side of Gion, and Sōkichi Takahashi in the Yase Amusement Park. Soon, though, the management of all the Shin Miura restaurants in Kyoto changed suddenly. Around 1929, Takahashi temporarily took over the Nijō Kawabata restaurant from Asa (Fig. 4.11). It was then closed around 1933, and a new restaurant managed by Ryōzō Shirai was opened in Kami-kiyamachi, to the north of Kiyamachi Oike. This was the first time that Shin Miura in Kyoto had crossed the Kamo River to the west, and once again it was in a *sekigashi* rental district with close associations with the *kagai*. Takahashi then took over the Gion branch from Shirai and started running the branch in addition to the Yase branch, a management system that was to continue until after the war.

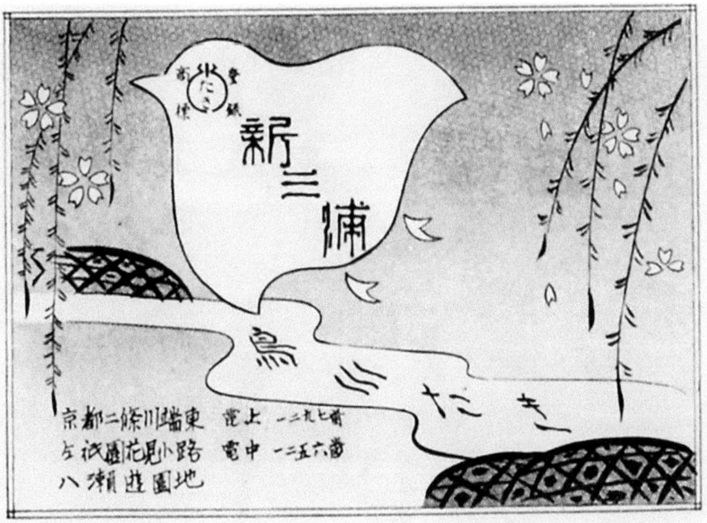

Figure 4.11 Advertisement with the locations of Shin Miura restaurants in Kyoto in the early Shōwa era (late 1920s)

Thus, the changes in the Shin Miura locations over time, with their focus on key entertainment spaces such as *kagai*, *sekigashi* rental districts, and suburban amusement parks, seems to have had a major impact on the culture of chicken cuisine and chicken restaurants in modern Kyoto.

4.4.2 Cloudy white soup

The "chicken *sukiyaki* cuisine" section in *Daikyōto* (1928) contains a row of chicken restaurants, including Junichirō Tanizaki's favored Kikusui. At that time, *sukiyaki* was the main type of chicken cuisine, but *mizudaki* was gradually gaining in popularity.

For example, Toriiwarō, an expansion of the Toriiwa chicken restaurant in Nishiki Market, promoted its "banquet of familiar chicken *sukiyaki* dishes that is becoming popular in the Eastern Capital [i.e., Tokyo] these days."[45] It had illuminated rooftop signboards on its northern and southern branches, both located in Saiseki, respectively advertising "chicken *kaiseki* and *mizudaki*" and "chicken *mizudaki* cuisine" (Fig. 4.12). A restaurant called Kinnabe, which sold pots made of pure gold, also sold *mizudaki*, as did chicken restaurants across the city.

Now, we previously mentioned differences in the spelling of *mizutaki* and *mizudaki* (see Section 4.3.1). The different spellings reflect the different ways the word has been written in the Japanese sources, but the descriptions we've read above also use "*mizutaki*" and "*mizudaki*" to refer to subtly different dishes.

Figure 4.12 Illuminated signboards of Toriiwarō

According to an acquaintance of the authors, in Hakata, the dish's home, it is generally called "*mizutaki*." Suigetsu, the dish's creator, claims to be "the birthplace of Hakata *mizutaki*."[46] Meanwhile, Shin Miura writes the word differently, advertising "Hakata specialty chicken *mizudaki*."[47] As we saw, proclaiming "the soup is good," Obiya distinguished between consommé- and potage-style soups, with Suigetsu representing the former and Shin Miura representing the latter. Apparently, the pronunciation *mizudaki*, with its voiced or "clouded" *da* consonant (the term for voiced consonant in Japanese literally means "cloudy consonant"), derives from the cloudiness of the soup served at Shin Miura.

The pronunciation used in Kyoto is of particular interest. Since as early as 1915, the term *mizudaki* has been used to describe the cloudy version of the dish. Toriiwarō, which opened after Shin Miura, also followed this nomenclature. Thus, the written and spoken term used in specialty chicken restaurants in Kyoto is now *mizudaki* (see Table 4.1 and Figure 4.13). Restaurants using kanji notation also employ the voiced consonant, and their soup is cloudy.

Table 4.1 Pronunciation and Notation of Mizutaki/Mizudaki in Kyoto

Restaurant	Location	Pronunciation	Japanese Notation
Toriiwarō	Nishijin	*mizudaki*	水だき
Shin Miura	Kami-Kiyamachi	*mizudaki*	水炊き
Torihatsu	Kawaramachi Aneyakōji	*mizudaki*	水だき
Toriyasu	Nawate-dōri	*mizudaki*	水だき
Torishin	Nawate-dōri	*mizudaki*	水だき
Toriyasa	Shimo-Kiyamachi	*mizudaki*	水炊き

Figure 4-13 Locations of chicken restaurants in Kyoto

Note: A black circle indicates a restaurant that is still in the same location, and a white circle indicates that the restaurant is no longer located there. See the final chapter for more information on the point marked with a star.

Torihatsu, a restaurant that reportedly served *sukiyaki* before World War II, "by following the pursuit of taste and continuous experimentation ... blended a creamy Hakata-style soup with a light Nagasaki-style broth" to create their signature style of *mizudaki* in the postwar era.[48] Toriyasu, a restaurant advertising itself as "specializing in a white chicken broth-based *mizudaki*" was reputedly established in 1955.[49] Torishin, located in an impressive *machiya* townhouse on the corner of Nawate-Shimbashi, initially focused on *sukiyaki*, shifting to *mizudaki* as their predominant dish after 1965.[50]

The modern history of chicken cuisine in Kyoto is thus marked by this transition from *sukiyaki* to *mizudaki*. This shift is perhaps best symbolized by Toriyasa, one of Kyoto's most venerable establishments, which was founded in 1788. As noted above, kabuki actor Kataoka Nizaemon XIII recognized Toriyasa as a famous "chicken *sukiyaki*" restaurant; so when did it start offering *mizudaki*? Below is a quote from an interview with the sixth-generation proprietor of Toriyasa:

> "Back then, the decking overhanging the river was larger than it is now, with two floors, an upper and lower deck, and they were full of *sukiyaki* customers," she said, recalling her childhood.
>
> "So in those days, it was chicken *sukiyaki*. When did you start offering *mizudaki*?"
>
> "That was after the war. My older sister, the fifth-generation owner; she was called Hatsuko Hirai. She lost her husband in 1946, and after that, she looked after the restaurant by herself. And she didn't just look after it, she actively researched chicken cuisine. She went out to Hakata in Kyushu and studied their local *mizudaki*, which she brought back and adopted at Toriyasa."[51]

Although Toriyasa had exclusively been a *sukiyaki* restaurant before the war, it became the specialty restaurant that it is now after the war, starting with *mizudaki*, which it learned from in Hakata. Rather than simply copying the Hakata version verbatim, Toriyasa took a creative approach to adopting *mizutaki* right from the start.

> I have always felt the energy that the town of Hakata possesses. We might call it a "Japanized" kind of energy. For example, when it comes to *zōni* [soup with mochi rice cakes], although Hakata *zōni* is said to be luxurious, Nagasaki *zōni* is an even more deluxe version. However, we can still see distinct traces of Nanban cuisine [dishes historically brought from outside Japan] in the Nagasaki *zōni*. Hakata *zōni*, however, has become a completely Japanese dish. The same is true of *mizutaki*, which has also become an excellent Japanese dish in Hakata. I believe that this excellence is something that Hakata can be proud of.[52]

It might perhaps be more appropriate to call this the "Hakata-ization" of dishes rather than "Japanization." Kyoto has also inherited this approach of taking dishes and making them its own.

> The original home of chicken *mizutaki* is Hakata. The origins of this cooking method appear to be Chinese, but since its arrival in Kyoto, everything from the tableware, the way the dish is eaten, the atmosphere, and so on, has been completely transformed into the Kyoto style.[53]

In this instance, Kyoto has adapted *mizutaki* in its own style, based on a dish that had in turn been transformed in Hakata. If tradition is one side of this shifting culture of cuisine and restaurants, the other side is undoubtedly creativity.

Notes

1. Tanizaki, Junichirō, "Suzaku Nikki (7)" (Suzaku diary, vol. 7), *Osaka Mainichi Shimbun*, May 4, 1900.
2. *Yomiuri Shimbun*, September 21, 1932.
3. Nihon Meisho Annaisha ed., *Sanfu oyobi kinkō meisho meibutsu annnai* (Guide to Kyoto, Tokyo, Osaka and the suburbs: Famous sights and locations), Nihon Meisho Annaisha, 1918, pp. 73–74.
4. *Tokyo Asahi Shimbun*, April 16, 1913; *Yomiuri Shimbun*, September 21, 1932.
5. Tanizaki, Junichirō, "Seishun monogatari" (The tale of youth) (*Tanizaki Junichirō zenshu, dai 13 kan* (The complete works of Junichirō Tanizaki, vol. 13), Chūōkōronsha, 1982), pp. 343–439. Quote from p. 400.
6. Kato, Masahiro and Kawasumi, Naomi, "Kindai Kyoto ni okeru shuyō shōtengai no tenpo fukugen: 'Gion-machi' o jirei toshita hōhō no kentō" (Restoration of the main shopping streets in modern Kyoto: Examining methods using the case of 'Gion-machi') (*Rekishi chirigaku* (Historical Geography), vol. 62 no. 4, 2020), pp. 1–17.
7. Nihon Shōkō Tsushinsha, ed., *Shokugyō betsu denwa meibo: Tokyo no bu, Taishō 11 nen ban* (Telephone directory by occupation, Tokyo section, 1922 edition), Nihon Shōkō Tsushinsha, 1922, p. 267.
8. Tanizaki, Junichirō, "Suzaku Nikki (4)" (Suzaku diary, vol. 4) (*Osaka Mainichi Shimbun*, April 30, 1900).
9. Tanizaki, Junichirō, "Kamigata no kuimono" (Food of Osaka and Kyoto) (*Tanizaki Junichirō zenshu, dai 22 kan* (The complete works of Junichirō Tanizaki, vol. 22), Chūōkōronsha, 1983 [first published in 1924]), 156–159 page. Quote from p. 159.
10. Ishida, Aritoshi, ed., *Kōshō gijyutsu: Miyako no sakigake* (Industry, commerce and technology: Pioneers of the capital), Ishida Ganjirō, 1883.
11. Kawaru, Tsūko, "Kyoto no inshoku ten" (Restaurants and bars of Kyoto) (Sasa Masaichi ed., *Yoru no Keihan* (Keihan at night) (*Bungeikai* vol. 16, Commemoration of regular special exhibition), Kinkōdō, 1903), pp. 63–89. All quotes below from pp. 64–65.
12. Makimura, Fumiaki, ed., *Osaka kotoba jiten* (The dictionary of the Osaka dialect), Kōdansha Gakubunko, 1984, p. 163.
13. Ōkubo, Tsuneji, *Shokutsū nyūmon* (Introduction to the way of the gourmet), Sōgensha, 1961, p. 68.
14. Nakagami, Naosaburo, ed., *Tairei kinen: Miyako odori shashin chō* (Enthronement ceremony memorial Miyako Odori photo album), Gigei Club, 1928.
15. Kataoka, Nizaemon, *Saga danwa* (Saga discussions), Sangatsu Shobō, 1976, p. 169.
16. Wakimoto, Zen'ichi, ed. *Hakata: Hiru to yoru no tabi (1962 nen ban)* (Journeys in Hakata: Night and day (1962 edition)), Hakata Shunjusha, 1962, p. 262. The quotes in this section are from the same book (pp. 262 and 264) unless otherwise noted.
17. Obiya, Einosuke, "Hakata no aji" (The taste of Hakata) (Editorial Department of Sōgensha, ed., *Kita-Kyushu mikaku chizu* (Kita-Kyushu Taste Map), Sōgensha, 1965), pp. 169–178. Quote from p. 169.

18. Fukumura, Kōji, *Zenkoku 50 toshi, naito rejyā annai: Aji, yado, asobidokoro de saguru miryoku no chihō toshi* (Nationwide 50 cities, night and leisure guide: Exploring attractive regional cities through flavors, accommodation, and entertainment), Jitsugyō no Nihonsha, 1970, p. 216.
19. Furukawa, Roppa, *Roppa Shokudan* (Roppa talks about food), Tokyo Sōgensha, 1955, p. 195.
20. Kiyohara, Iseo, ed., *Fukuoka shi* (Fukuoka city), Fukuoka City Editorial Department, 1916, pp. 294–295.
21. Kiyohara, *Fukuoka shi*, pp. 298–299.
22. Here, using the "Chronological topographic map viewing site 'Maps of then and now on the web'" (https://ktgis.net/kjmapw/) developed and maintained by Professor Kenji Tani (Saitama University), with reference to 1887 (Meiji 20) Rapid Survey Map "Kokura Vicinity" (1:50,000) and Taisho 15 (1926) Survey Map "Fukuoka" (1:25,000).
23. Mizobe, Nobutaka, ed., *Fukuoka shi annai ki* (Fukuoka city guidebook), Sekizenkan Shiten, 1910, p. 119.
24. Matsukawa, Jirō, *Zenkoku kagai meguri* (Nationwide tour of entertainment districts), Seibundō, 1929, p. 584.
25. Matsukawa, *Zenkoku kagai meguri*, p. 588.
26. Mizobe, *Fukuoka shi annai ki*, p. 117.
27. Matsukawa, Jirō, *Kanraku kyō meguri* (Tour of nightlife districts), Santokusha, 1922, p. 355. Reference was made to the reprinted edition (2016) by Kastori Shuppan.
28. Obiya, "Hakata no aji," p. 169.
29. Referred to Suigetsu's website (http://www.suigetsu.co.jp/hist.html), last viewed May 29, 2021.
30. Obiya, Einosuke, *Karā bukkusu 436: Hakata no aji* (Color books 436: The taste of Hakata), Hoikusha, 1978, p. 127.
31. Kiyohara, *Fukuoka shi*, p. 290.
32. Yagito, Shigeo, ed., *Chikuzen: Hakata*, Fukuoka Kyōwakai, 1938, p. 27.
33. *Asahi Shimbun*, January 15, 2006.
34. Ueno, Masao, ed., *Kyushu shinshi roku, dai 1 shū* (Who's who of Kyushu vol. 1), Shūhōsha, 1914, p. 219. Imaizumi Kenzō, *Fukuoka shi shō kō jinmei roku* (Directory of merchants and craftspeople in Fukuoka city), Hakata Chamber of Commerce, 1918, p. 161.
35. Kokubu, Ayako, *Zoku, Kyō no okami tachi: Shinise no aji o tazunete* (Continued: Kyoto hostesses: Visiting the long-established restaurant), Shibata Shoten, 1984, p. 74.
36. Her name was first confirmed with regard to Kyoto in Kyoto Central Telephone Bureau, ed. *Kyoto denwa bangō bo* (Kyoto Telephone Directory), 1919.
37. Sōgensha Editorial Department, ed., *Kyoto mikaku chizu: 1969 nen ban* (Kyoto taste map 1969 edition), Sōgensha, 1968, p. 85. The author of this entry is Jinnosuke Inoue.
38. Takkeinomon, Shizue, "Kyoto no kagai" (Kyoto entertainment districts) (Sasa Masaichi ed., *Yoru no Keihan* (*Bungeikai* No. 16, Commemoration of regular special exhibition), Kinkōdō, 1903), pp. 26–63. Quote is from pp. 37–38.
39. Terao, Kōji, "Ima wa mukashi, mukashi wa ima: Bunkyō chiku ni tsuite" (The present is past, the past is present: On educational districts) (*Rakumi* vol. 70, 1957), pp. 37–40. Quote from p. 39.

40. Terao, "Ima wa mukashi," p. 39.
41. Sōgensha Editorial Department, *Kyoto mikaku chizu: 1969*, p. 84.
42. Matsuo, Otojirō et al., eds., *Kyoto shō kō jinmei roku; Taishō 15 nen kaiban* (Records of famous Kyoto merchants: Revised 1926 edition), Kyoto Shōkō Jinroku Hakkōjo, 1925. Kyoto Central Telephone Bureau, ed. *Kyoto denwa bangō bo: Taishō 15 nen 8 gatsu tsuitachi genzai* (Kyoto telephone directory, as of August 1, 1926), Kyoto Central Telephone Bureau, 1926.
43. Kōjunsha ed., *Taishō 15 nen yō, dai 29 ban, Nihon shinshi roku* (Who's who of Japan, 1926, 29th edition), Kōjunsha, 1925, Fukuoka, p. 44.
44. Kokubu, *Zoku, Kyō no okami tachi*, p. 78.
45. Nitta, Yoshihiko, ed., *Kyoto gadan taikan: Fukakujyuku jyukuin meibo* (Overview of the art world of Kyoto: List of school members), Toshi to Geijutsusha, 1931.
46. Suigetsu website (http://www.suigetsu.co.jp/index.html), last viewed on August 14, 2021.
47. Shin Miura Hakata Honten website (http://www.Shin Miura-honten.com/index.html), last viewed on August 14, 2021.
48. Torihatsu website (http://www.kyoto-torihatsu.com/history.html), last viewed on August 14, 2021.
49. Toriyasu website (http://toriyasu.net/index.php?InformationPage), last viewed on August 14, 2021.
50. Torishin website (http://torishin.my.coocan.jp/sub1.htm), last viewed on August 14, 2021.
51. Kokubu, Ayako, *Kyō no okami tachi: Shinise no aji o tazunete* (Kyoto hostesses: Visiting the long-established restaurant) Shibata Shoten, 1980, pp. 119–120.
52. Obiya, "Hakata no aji," p. 169.
53. Kokubu, Ayako, "Toriyasa: Tori mizutaki" (Toriyasa: Chicken mizutaki) (Sōgensha Editorial Department, ed., *Kyoto mikaku chizu: 1969 nen ban*, Sōgensha, 1968), pp. 128–129. Quote from p. 129. Kokubu also commented on Shin Miura: "The Hakata specialty, chicken *mizudaki*, is the signature dish, but the location, cooking, seasoning, and ingredients have somehow incorporated Kyoto-like characteristics, making it a completely Kyoto-style flavor. The flavor is a perfect match for the atmosphere of Kyoto." Kokubu, Ayako (text), Hamabe, Kiyoji (photograph), *Kamera Kyō aji 100 sen* (100 selections of Kyoto flavors: Camera edition), Tankō Shinsha, 1963, p. 28.

10.2478/9784991351600-012

Chapter 5
Western-style Cuisine in the *Kagai*

5.1 The Scent of the Early Shōwa Period

5.1.1 Western-style food once a day

During his stay in Kyoto in 1960, Shōtarō Ikenami received the Naoki Prize, which is awarded for the best work of popular literature by a rising author in Japan. More than a decade later, looking back on the time when he received the award, he recalls that "I just didn't feel satisfied if I didn't eat [Takarafune]'s Western-style food once a day."

> [Takarafune] is located on the northern side of Gion, west of Hanamikōji street, and it still has the scent of Western-style cuisine from the early Shōwa period. I feel nostalgic for it, so I go there to eat.
>
> I heard that it was founded at the end of the Taishō era [1912–1926] or the beginning of the Shōwa era [1926–1989].
>
> Its atmosphere is very much that of a Western-style restaurant in the Gion district, with geisha making an appearance during the day to eat croquettes and spaghetti. I loved their beef and veggie sandwiches, cutlets, and Indian chicken.
>
> I wonder if they're still doing it, the Indian chicken.
>
> It is a dish of buttered rice topped with a soft-boiled egg and chicken simmered in curry sauce.[1]

Takarafune (more precisely, Grill Takarafune), which opened in 1927, was the oldest Western-style restaurant in the Gion Kagai district. The name of the restaurant was written in kanji (宝船) in the prewar Shōwa period, and later written using a mixture of kanji and hiragana (たから船). Sadly, Takarafune, with its "scent of Western-style cuisine from the early Shōwa period" and its atmosphere which was "very much that of a Western-style restaurant in the Gion district," is now long closed.

The owner of Takarafune at the time Ikenami was visiting stated that "our restaurant offers something different from the 'very much in vogue' French cuisine"[2] and that "our menu doesn't change; we keep doing the same thing we've been doing for ages."[3] This culinary philosophy seems to have appealed to Ikenami, who said he went there "once a day." Indeed, on days when he wasn't able to go to the restaurant because he was writing a novel or screenplay, his devotion to Takarafune was such that he would have food from there delivered to his lodgings, as we can see from his diary of July 11: "for lunch, I had cold consommé and a beef sandwich, which was delivered to my lodgings" from Takarafune.[4]

"Beef sandwich" may refer to the "beef cutlet sandwich." An article on Takarafune indicates that the beef cutlet sandwich was also popular as a local product. Ikename describes a beef cutlet sandwiched between crispy, thick slices of toasted bread "with delicious sauce dripping from it. Although it's definitely good hot, it's also great if you eat it the next day. Such are the reviews from devotees of the Takarafune sandwich."[5]

5.1.2 Chicken rather than beef

Table 5.1 below is a reproduction of Western-style menus with prices from two restaurants from the period when Takarafune was founded. For comparison, a summary of the menus and prices for Japanese-style food, drinks, and sweets at the same two restaurants is provided in Table 5.2: Daimaru Gōfukuten, located in Shijō Takakura, and Bussankan, located in front of Kyoto Station. The former opened in October 1928 in a new eight-story building, with the dining room on the sixth floor. The latter was rebuilt in October 1926 and changed its name from Kyoto Bussankan to Bussankan.

Table 5.1 Western-style menus from the early Shōwa period (1926–1928)

Daimaru Gōfukuten (Shijō Takakura)	Bussankan (in front of Kyoto Station)
Takeaway meal (four dishes with coffee) (1 yen)	Set meal (1 yen)
Lunch (two dishes with rice and coffee) (50 sen)	Bussan lunch (with coffee) (50 sen)
Soup (30 sen)	Soup (30 sen)
Fried shrimp (50 sen)	Fried shrimp (50 sen)
Croquettes (30 sen)	Croquettes (35 sen)
Chicken cutlet (55 sen)	Chicken cutlet (50 sen)
Cutlet [beef] (40 sen)	Beef cutlet (35 sen)
Cutlet [pork] (40 sen)	Beef steak (50 sen)
Beef steak [grilled] (70 sen)	Chicken sautee (50 sen)
Cold meat (50 sen)	Mince balls (35 sen)
Omelet (35 sen)	Omelet (30 sen)
Chicken rice (40 sen)	Ham and egg (40 sen)
Hayashi rice [hashed beef rice] (40 sen)	Chicken rice (with coffee) (40 sen)
Curry rice (35 sen)	Haishi [Hayashi] rice (with coffee) (35 sen)
Sandwich (40 sen)	Curry rice (with coffee) (35 sen)
Vegetable salad (25 sen)	Sandwich (40 sen)
	Vegetable salad (25 sen)
	Ham salad (40 sen)
	Asparagus (35 sen)

Table 5.2 Menus for Japanese-style food, drinks, and sweets from the early Shōwa period (1926–1928)

Daimaru Gōfukuten (Shijō Takakura)	Bussankan (in front of Kyoto Station)
Japanese food	Japanese food
Takeaway meal (1 yen)	Kaiseki (80 sen)
Makunouchi bento (50 sen)	Makunouchi (50 sen)
Unagidon (60 sen)	Unagidon (60 sen)
Oyakodon (40 sen)	Oyakodon (40 sen)
Tokyo sushi (40 sen)	Sushi (40 sen)
Chirashi [sushi scattered over a bed of rice] (40 sen)	Chirashizushi [same as "chirashi"] (40 sen)
O-suimono [soup] (25 sen)	O-suimono (20 sen)
Kimosui [eel liver soup] (25 sen)	Kimosui (10 sen)
Shiruko [adzuki bean soup with mochi rice cake] (15 sen)	Shiruko (10 sen)
Strawberry ice (10 sen)	Strawberry ice (15 sen)
Drinks and sweets	Drinks and sweets
Coffee (10 sen)	Coffee (10 sen)
Black tea (10 sen)	Black tea (10 sen)
Lemon squash (25 sen)	Lemon squash (25 sen)
Cider [apple juice] (25 sen)	Cider [apple juice] (25 sen)
Calpis (15 sen)	Calpis [soft drink] (20 sen)
Soda water (10 sen)	Soda water (15 sen)
Western sweets (15 sen)	Western sweets (10 sen)
Jelly (15 sen)	Jelly (15 sen)

> Our dining room is located on the fifth floor, where the knives [of our chefs] prepare Hasshin's finest Japanese and Western cuisine. We hope that you will come and enjoy our offerings at your leisure, as you listen to our wireless telephone [i.e., a radio].[6]

"Hasshin" here may be referring to the long-standing establishment Hasshintei located in Fuyacho Oike (later at the corner of Yanagibaba and Oike streets).

Looking at the two tables, we can see that both the "takeaway price meal" (*nakashoku*) and "set meal" (*teishoku*) cost 1 yen, with lunch for 50 sen (Table 5.1), the latter being the same as the *makunouchi bento* (boxed lunch containing rice and several different side dishes), which was cheaper than *unagidon* (eel over rice) but more expensive than *oyakodon* (chicken and egg over rice) (Table 5.2). It is unclear what the set meal included, but it was probably an assortment of dishes from the à la carte menu.

So, we have staple dishes of Japanese Western-style cuisine, such as croquettes, cutlets, and fried shrimp. The pricing of the dishes is noteworthy, with chicken being more expensive than beef. Although the beef steak at Daimaru is the most expensive single dish, the beef steak and chicken sautee at Bussankan are the same price. Chicken was clearly in high demand.

Tables 5.1 and 5.2 are based on information from *Daikyōto*, a guidebook published to coincide with the ascension of Emperor Shōwa (Emperor Hirohito). Interestingly, among the preparations for the emperor's arrival in Kyoto, "a poultry breeding farm has been especially established at the Prefectural Agricultural Experiment Station in the scenic area in northern Kyoto," with the "Rhode Island Red" for meat and the "Nagoya breed" for eggs. As the chicken was to be "served to Their Imperial Majesties, the Emperor and Empress," chickens that had been through "many rounds of careful selection" were ordered.[7] This truly was an era of chicken rather than beef.

Going back to the Western-style menu, other meat dishes include "cold meat" (Daimaru) and "mince balls" (Bussankan). Where there is chicken, there are also eggs; hence, we also find "ham and egg" and omelet dishes on the menu.

The menus also include light meals that are standard at any modern-day coffee shop, such as chicken rice, *hayashi* rice, curry rice, sandwiches, and salads. The modern reader may be surprised to note that lemon squash, cider, and Calpis (a soft drink, at Bussankan) are at least double the price of coffee.

The "Department store and station cafeteria" section in *Daikyōto* doesn't have a "Western-style cuisine" section; instead, "Western-style food" and "Japanese-style food" are written alongside each other, suggesting that Japanese style "Western cuisine" was already established as a genre in Kyoto when Takarafune opened.

5.2 The development of Western-style chain restaurants

5.2.1 Western-style cuisine is everywhere

The reason Ikenami, after the war, went out of his way to seek the "scent of Western-style cuisine from the early Shōwa period" was probably because the food culture that had been accepted as Western food in the Meiji period had become established as Japanese Western-style food by the beginning of the Shōwa period. It may also have been because this particular take on Western-style food gradually evolved into something else after the war. When Takarafune opened, there were already many Western-style restaurants operating in Kyoto (Table 5.3). But Western-style restaurants did not begin to spread in Kyoto until the late Meiji period, so their history in the city is by no means a long one. At the turn of the twentieth century, its presence was widespread, although the reviews were somewhat ambivalent:

> Western-style cuisine is everywhere, and even the local beef restaurant will take orders for cutlets, stews, and beef steaks. Of course, Western-style food at hotels catering to foreigners is perfectly good, but there are many other commonplace Western-style restaurants that incite little enthusiasm.[8]

Table 5.3 Western-style restaurants in the early Shōwa period (1928)

Restaurant Name	Location
Manyōken	Eastern Shijō Fuya-chō
Yaomasa	Western end of Shijō Ōhashi Bridge
Yaomasa Kitaten (Northern Branch)	Western end of Shijō Ōhashi Bridge
Kikusuikan Honten (Main Branch)	Eastern end of Shijō Ōhashi Bridge
Kikusuikan Shiten (Branch)	Corner of Shijō Teramachi
Tōyōtei Honten (Main)	Southern Kawaramachi Sanjō
Tōyōtei Shiten (Branch)	Eastern Shijō Yamato-ōji
Nōen	Eastern Shijō Nawate
Ichiyōken	Northern Higashi Kiyamachi Shijō
San'yōken	Southern Yamato-ōji Shijō
Fukuya	5-Chome Miyagawasuji
Kaiyōtei	South of the stone steps of Yasaka Shrine, Gion
Kaiyōtei Shiten (Branch)	Southern Karasuma Shijō
Edo Café	Northeastern Shinkyōgoku Shijō
Keihan Shokudō	Southern Nawate Sanjō
Kaiseytei	Northern Karasuma Matsubara
Star Shokudō	In front of Nishiki Tenmangu Shrine, Shinkyōgoku
Murase Shokudō	Northern Teramachi Nishikikōji
Salon Komadori	Eastern Shijō Yanaginobanba

However, when Junichirō Tanizaki visited Kyoto in the spring of 1912, the final year of the Meiji era, he was rather impressed by the spread of Western restaurants.

> I was guided to a Western-style restaurant called Bukkoku Ryōri Manyōken [French Cuisine Manyōken] in Fuya-chō. It seems that Western-style food in Kyoto has developed in a short time in recent years, with cafes and even *paulista* [Brazillian-style coffee shops] setting up shop. This restaurant opened recently in a renovated doctor's residence, which gives it its characteristic appearance, but it is said to have a good reputation. The building is in fact Japanese in style, and the tatami are covered with [Western-style] rugs, tables, and chairs. The inner garden of the restaurant, about five *tsubo* [16 square meters] in area, is covered in green moss, and the antique patina lent by the stone lanterns and other such items is somewhat wasted in a Western-style restaurant.[9]

As indicated by the phrase "Western-style food in Kyoto has developed in a short time," the number of restaurants specializing in Western cuisine increased. Up to that point, accommodation facilities had doubled up as restaurants,[10] and it was not until the late Meiji period that specialty restaurants appeared.

Tanizaki was describing lunch at Manyōken, which had opened not long before in 1910 and which still operates today, one of the oldest Western-style restaurants in Kyoto. Its style anticipates the "Machiya French" that became popular in the 2000s, with features including a repurposed "doctor's residence" with tables and chairs placed on tatami mats.

Among the Western-style restaurants that opened from the late Meiji period to the Taishō period, it is worth noting that those shown in bold in Table 5.3 are still in business today, albeit with considerable changes to their locations and business conditions. Having shed themselves of their initial status of "[inciting] little enthusiasm," these Western-style restaurants have become firmly entwined in the roots of Kyoto's catering industry.

5.2.2 Selecting and developing locations

By one account, Tōyōtei was opened by Ginjirō Takahashi as a Western-style restaurant in 1901.

> Western-Style Food Business: The owner Ginjirō Takahashi was born in 1870 in Katori, Ibaraki Prefecture. As a young man, he came to Kyoto and learned the trade as a chef at San'yōtei, a western restaurant. He later moved to the Osaka Hotel, where he worked for three years, and then further refined his skills by working as a chef for C. Illies & Co. in Kobe, before opening an establishment at the current location in 1901. His devoted efforts yielded great results and a new company office was built in 1915.[11]

Although other accounts indicate that the restaurant might have originally opened a few years earlier, either way, Tōyōtei is among the oldest of the Western-style restaurants in Kyoto today. Its importance in the history of Western-style cuisine in Kyoto is partly due to its longevity, but perhaps more important is the distinctive approach it took to selecting branch restaurant locations as it developed into a chain.

Table 5.4 shows the branches of Tōyōtei, Manyōken, and a newcomer in among the Western-style restaurants Star Shokudō, which rapidly opened a chain of new restaurants in the Shōwa period (1926–1989). As indicated in Table 5.3, Yaomasa had a branch on the north side of Shijō-dōri, Kikusuikan had a branch at the corner of the intersection with Teramachi-dōri facing Shijo-dōri, and Kaiyōtei had a branch in southern Karasuma Shijō.[12] It was in this context that each of the three restaurants shown in Table 5.4 had three or more branches.

Although Tōyōtei and Manyōken abolished their branches during and after the war, starting with its main restaurant in northern Teramachi Nishiki, Star Shokudō expanded from the five restaurants in Table 5.4 to eight restaurants between the late 1950s and early 1960s. Their locations included: Kyōgoku (northern Shinkyōgoku Shijō); Gion (eastern Shijō Nawate); Kawaramachi (northern Kawaramachi Shijō);

Nishijin (northern Senbon Ichijō); Ōmiya (eastern Shijō Ōmiya); Tenbōkaku (observatory at Mount Hiei); Yase (within the Yase Amusement Park); and Hikone (Shiga Prefecture).[13] Two notable branches here are the one in Yase Amusement Park, discussed in the introduction; and the summit of Mount Hiei, as if they were hot on the heels of their competitor Manyōken.

Table 5.4 Western-style chain restaurant developments (early Shōwa period)

Restaurant	Location of Main Restaurant	Branch Restaurants
Tōyōtei	Southern Kawaramachi Sanjō	1. Gion 2. Kawaramachi Marutamachi 3. Nijō Teramachi 4. Sanjō Shinmachi 5. In front of Shimabara Ōmon Gate 6. Kawaramachi Bukkoji Temple
Manyōken	Eastern Shijō Fuya-chō	1. Daimai Shokudō 2. Shimeidake Shokudō 3. Yase Shokudō
Star Shokudō	Northern Teramachi Nishiki	1. Star Bar (Shinkyōgoku Nishiki) 2. Star Soda Fountain (northern Shinkyōgoku Shijō) 3. Star Shokudō Nishijin Shibun (northern Senbon Nakadachiuri) 4. Star Shokudō Demachi Shibun (northern Kawaramachi Imadegawa) 5. Gion Star Soda Fountain (eastern Shijō Nawate)

The locations of the Manyōken branches differ somewhat depending on the advertisements one refers to. For example, a newspaper advertisement in 1926 lists the "Shimeidake Shucchōten" (Shimeidake Branch) and "Yase Shucchōten" (Yase Branch), as well as two others, the "Minamiza Shucchōten" and "Mayville Shokudō" (on the corner of Sanjō and Nishinotōin).[14] The latter was probably located in the Meiji Bank branch building on the northwest corner of Shijō Higashinotōin.

However, a newspaper advertisement from 1929 mentions only "Shimeidake Shokudō" and "Yase Shokudō." The Shimeidake branch was located on the upper floor of the station building of the Shimeidake Cable Car Station and is advertised as having "seating for 300 guests." The Yase branch was similarly positioned as a "convenient cafeteria for the people."[15]

Returning to Table 5.4, "Daimai Shokudō" was a branch of Manyōken located in the basement of the former Kyoto branch office of the Osaka Mainichi Shimbun, which is now known as the "1928 Building," built in 1928 with a distinctive Art Deco style, on Sanjō-dōri. At the time that Junichirō Tanizaki visited Manyōken, it was located in Umeya-chō, northern Fuyachō Nishikikōji. Tanizaki describes its location as "on the wide street of Shijō, where train lines are being expanded due to the revision of the city wards,"[16] referring to when Manyōken moved to Naramono-chō in eastern Shijō Fuyachō, taking advantage of the expansion of Shijō-dōri (in what is referred to as the "revision of city wards" in Tokyo), which was part of three major development projects in Kyoto at the time (digging the Lake Biwa Canal, developing the water supply and sewerage, and widening roads to install streetcar tracks). Following the advice of Zensuke Ōsawa of the Kyoto Electric Railway, the new Manyōken main restaurant on Shijō-dōri was designed by authoritative architect Goichi Takeda and opened in 1913. The Kyoto branch office of the Osaka Mainichi Shimbun (now the 1928 building) was also designed by Takeda.

Takeda named the saloon in the main restaurant "Ranteibō" using kanji in a pun on the French word "rendezvous." Given that this name was also adopted as the title of the company history commemorating the 75th anniversary of Manyōken,[17] it may be that the "Omai Shokudō" branch was established through the connection with Takeda.

In April 1912, having alighted at Kyoto Station, Tanizaki headed for the Kyoto branch office of the *Osaka Mainichi Shimbun*. The record of his stay in Kyoto from April 27 to May 28, the "Suzaku Diary," was serialized for 19 issues in the *Osaka Mainichi Shimbun*. Tanizaki's hosts took him to Manyōken for lunch immediately after he arrived. The main restaurant was relocated the year following Tanizaki's visit to a new building designed by Takeda, who as we just saw was later involved in the design of the Kyoto office of the *Osaka Mainichi Shimbun*, with a Manyōken branch situated in its basement. Here, we see a consequential connection between a writer, an architect, a newspaper publisher, and a chain of Western-style restaurants.

Then, we have Tōyōtei, which had six branches as of 1930. One notable thing about Tōyōtei is the Kawaramachi-dōri location. When the main restaurant opened in the street in 1901, Kawaramachi-dōri was a mere backstreet.

However, when widening work was completed to install a streetcar line between Marutamachi and Shijō in 1926 and between Shijō and Shichijō in 1927 Kawaramachi-dōri became a main street and was expected to develop into an upmarket shopping area along the lines of Tokyo's Ginza district.[18] Tōyōtei opened a new branch on the widened Kawaramachi-dōri, immediately adjacent to the main restaurant, one building literally supporting the other. The restaurant at Nijō Teramachi was situated in a hotspot until the Shijō Kawaramachi intersection was created, so choosing to build restaurants in both locations struck a good balance between old and new.

It is also important to note that there were branches of Tōyōtei in Gion-machi and in front of Shimabara Ōmon Gate. The former is on the part of Shijō-dōri that runs through central Gion Shinchi Kōbu from east to west, and the latter is in front of the historic Ōmon Gate in Shimabara, and both had close connections with the main *kagai* entertainment districts of modern Kyoto.

5.2.3 Restaurants and cafes of Gion

The restaurant mentioned at the beginning of this chapter, Takarafune, first opened in 1927 in premises on the western side of Katoriya, a long-established bag shop (now designated a building of historical significance), on the south side of the Gion district facing Shijō-dōri. After the war it moved to a backstreet on the north side of Gion (Tominaga-chō). So let us explore the streetscape of the Gion district when Takarafune was founded.[19]

Walking along Shijō-dōri from the bottom of the stone steps of Yasaka Shrine toward Shijo Ōhashi Bridge, Suifu Kishimoto (1892–1965), renowned for his *senryū* poetry (a form similar to *haiku*), describes "walking along looking at the lights of unique restaurants such as Jūnidanya and Hiranoya, as well as cafés such as Tōyōtei and Nōen, and before long seeing the tower of Minamiza theater."[20] In Suifu's eyes, the streets of the early Shōwa period were colored not only by the unique restaurants but also the cafés, which were a new and burgeoning form of service business at the time. According to a geisha of the Gion Shinchi Kōbu district, speaking in the distinctive Kyoto dialect at the time, "The first café in Gion was Tōyōtei. The next was Nōen."[21] These were also the two cafés mentioned by Suifu and can be regarded as the earliest of such establishments in the Gion district.

Hikaru Saitō's *The Age of the Phantom Café* has more information on modern Kyoto cafés. According to Saito, the Gion Café opened in November 1912 and Nōen (meaning "garden" or "farm") opened in the summer of 1913:[22] Shūji Yamamoto similarly recalls "If memory serves me well, I believe that the first café in Kyoto was Gion Café, whose name still lives on in the *ishidanshita* area [below the stone steps of Yasaka Shrine]. At that time, Tōyōtei was still a dingy one-story affair."[23] Likewise, Kenzō Maeshiba recounts that "Tōyōtei's 'Gion Café' was located on the south side of Hanamikōji, a little to the west," so Gion Café must have made an impression on him, worthy of multiple mentions.[24]

When Gion Café, the Gion branch of Tōyōtei, opened, the founder, Ginjiro Takahashi, said, "I invented a special seat where geishas could sit and feel comfortable,"[25] indicating his intention that geishas would visit, taking advantage of its location on the central north–south axis of Gion Shinchi. Ikenami, catching glances of geishas eating croquettes and spaghetti in Takarafune, had a sense of its "atmosphere [being] very much that of a Western-style restaurant in the Gion district," and the Gion branch of Tōyōtei was truly at the forefront of Western-style establishments that catered to the *kagai* entertainment districts. In fact, in the first decade (1926–1936) of the Shōwa period, the restaurant began to advertise that "the spirit of the new Gion has come to Gion Tōyōtei Grill."[26]

Roppa Furukawa (1903–1961) soon pointed out this characteristic of Western-style establishments located in and around *kagai*, naming them "*iromachi yōshoku*," where *yōshoku* literally means "Western-style food" and iromachi is a local term for the "pleasure districts."

5.3 An extension of the *kagai* entertainment districts

5.3.1 Roppa Furukawa's theory of Western-style food

According to Tsuneji Ōkubo:

> Roppa Furukawa's concept of Western-style food catering specifically to the pleasure districts is right on the mark. Simply put, the Western-style food of the pleasure districts was the most "Japanese" Western-style food. When Western-style food was still a luxury, *Western-style restaurants opened exclusively to cater to the extravagant tastes of pleasure district punters, where the seasonings and other aspects of the dishes were specially prepared to appease customers who were unfamiliar with Western food.* You can still inadvertently bump into this "*iromachi yōshoku*" in Tokyo, Kyoto, and Osaka, albeit under a new katakana name.[27]

Elsewhere, he states:

> Western-style food in Osaka was first available at hotels, but Western-style restaurants also opened in the *karyūgai* ["flower and willow districts," i.e., pleasure districts], where the geisha are. Western food, which was considered luxurious at the time, was only available in the *iromachi* (pleasure districts), where a Japanese-style of Western food called "*iromachi yōshoku*" was created.[28]

Meanwhile, Roppa admits that "my self-indulgent pleasures have been accompanied by the Western-style food of the various *kagai*,"[29] before going on to say:

> However, I myself have never used clever words like "*iromachi yōshoku*." Mr. Ōkubo has skillfully rephrased what I was calling "Japanese-style Western food." However, the phrase "*iromachi yōshoku*" does have a certain ring to it.[30]

Roppa's essays on food, titled "Roppa Shokudan" (roughly, "Roppa talks about food"), was serialized in *Amakara* magazine from its 18th issue in February 1953.[31] Therein, Roppa discusses the topic of Western food six times in a row, and although he repeatedly mentions "Japanese-style Western food" and "Japanized Western food," he did not use the term "*iromachi yōshoku.*"[32]

The first edition of Ōkubo's *Pilgrimage of Delicious Things* was published in 1956, and is where he first uses the phrase "*iromachi yōshoku.*" Thus there is no doubt that he coined the phrase. Although Ōkubo looks down on *iromachi yōshoku* as being "specially prepared to appease customers who were unfamiliar with Western food," Roppa is more positive, acknowledging that the phrase '*iromachi yōshoku*' does have a certain ring to it." Tokyo-born Roppa also does not overlook differences in language: "However, this phrase only works in Kansai, because he doesn't use the word '*iromachi*.'"[33]

It is indeed true that in the Kyoto and Osaka area, the *kagai* are referred to either as "*kuruwa*" or "*iromachi*." Roppa continues, "It is certainly not in Tokyo, nor is it in Osaka, but rather in Kyoto, that the term '*iromachi yōshoku*' fits best."[34]

In Kyoto, he finds a distinct characteristic of *iromachi yōshoku*.

> Is it not the case that upon first entering places such as San'yōken in Gion and Ichiyōken in Kiyamachi (there are so many restaurants in Kyoto called "such-and-such-yōken"), one sees a sight that cannot be seen in other places in terms of the building or decoration? There are curtains set up so that customers cannot see each other. So, no matter which table you sit at, the curtains will be drawn immediately...
>
> ...In other words, rather than "restaurants" [here the original text uses the English word "restaurant"], these Western-style establishments are an extension of the pleasure districts and *iromachi*.
>
> For that reason, in places like this, there's always a boy or some elderly fellow who will bend over backward for you.[35]

The most significant feature of *iromachi yōshoku* that Roppa had gleaned from his experiences of self-indulgent pleasures over the years was that Western-style restaurants were an extension of the *kagai*.

5.3.2 Restaurant sub-branches and delivery service

Roppa observes that "there are so many restaurants in Kyoto called 'such-and-such-yōken,'' and indeed we do find multiple restaurants whose names end with "-yōken" (養軒 in kanji, which loosely translates as "house of nourishment"). These restaurants also tend to be in and around the *kagai* entertainment districts. The restaurant that Roppa mentions, Ichiyōken, is located back-to-back with Pontochō (in northern Kiyamachi Shijō), whereas San'yōken is close to Gion Shinchi Kōbu (in northern Yamato-ōji Shijō). There was also a Goyōken, which we will see later, located on Hanamikōji Street on the south side of Gion. The Kamishichiken entertainment district, which opened in front of the gates of Kitano Tenmangū Shrine, also had a restaurant named Hokuyōken.

However, not all Western-style restaurants in Kyoto featured the "yō" kanji (養- meaning "nourishment") in their names. There was of course Takarafune in Gion, which opened in 1927, as well as Kaiyōtei (the "yō" here is a different kanji 陽, meaning "sun" or "positivity"), which opened in 1917 at the eastern end of the area below the stone steps of Yasaka Shrine.[36]

> That's right, before the war, Kaiyōtei was located on a street with trains, a short way up from the area below the stone steps; it now has a restaurant on Kiyamachi. The Western-style food here really has that Taishō-era flavor, and I think at that time, it was the only place in the Gion neighborhood where the French names of the dishes were on the menu. It was in this old-style French establishment that geisha and their patrons, rendered quite useless by one too many geisha parties, would lie around until late in the night.[37]

Thus, in contrast to Shōtarō Ikenami (1923–1990), who kept seeking the "scent of Western-style cuisine from the early Shōwa period," Yoshikata Yoda (1909–1991) longs for "that Taishō-era flavor." Both writers yearned for the taste of their childhood. It seems that the *iromachi yōshoku* Western-style restaurants were an extension of the *kagai*, where the drunken guests from geisha parties at the teahouses could idle away the hours.

Grill Fukuya, situated in the Miyagawa-chō *kagai*, is another restaurant with a long history that is still in business to this day. According to one source, Fujiya (Grill Fujiya) opened in 1913 in Shimogawara, a former *kagai* in front of the torii gate of Yasaka Shrine.

> ...it first opened in 1913 [Taishō 2]. At that time, studying Western cuisine at a Kyoto hotel meant nothing less than going through full-blown French training. Some of the customers were keen to eat Torakichi's cuisine outside of the hotel, so he secretly opened a restaurant that provided a place for familiar customers to eat his cuisine, while continuing to work at the hotel. Due to its proximity to the *kagai* of Gion, the stylish Western cuisine that Torakichi produced quickly developed a good reputation among geisha and maiko. We can gain a clear idea of just how popular this restaurant was at the time as it was said that the restaurant was able to break even just with the delivery orders from the patrons of geisha teahouses.[38]

However, the *Asahi Shimbun* indicates that it was a few years older than that:

> Although the image of the *kagai* has a strong association with Japanese-style food, Western-style food has also been much liked there for many years. In Miyagawa-chō, Higashiyama Ward, one of the five *kagai*, there is a Western-style restaurant, Grill Fukuya, which was founded 110 years ago. The restaurant is always full of fans of Kyoto-style food [*kyōmaru-uchiwa*] that geisha distribute to their clients, bearing the names of the geisha and maiko that they frequent.
>
> The specialty, "Fukuya rice," was created around 80 years ago at the request of a geisha who wanted "a rice omelet with more embellishments."[39]

The newspaper claims that 2017 was the 110th anniversary of Fukuya, which means that it opened around 1907 in the late Meiji era. In fact, "Fukuya" (on the north side of the eastern end of Matsubara Ōhashi Bridge) is listed in the "Western-style cuisine business" section under "Kabuto beer hall, Western-style cuisine, bread" in

1912 (“Kabuto” here referring to a brand of beer).[40] It is one of the oldest *iromachi yōshoku* restaurants in Kyoto.

An advertisement for Fukuya (currently Grill Fukuya) from the subsequent Taishō period (Fig. 5.1) indicates that the restaurant had a branch in the shopping district on Teramachi-dōri, as well as a sub-branch located inside the Kaburenjō theater in the Miyagawa-chō *kagai* that was only open during special performances. Thus, its connection with the *kagai* was more than simply an extension. This was also the case with Ichiyōken, located in Kiyamachi, which opened a sub-branch in the basement of the Kaburenjō theater in Pontochō during the Kamogawa Odori dance performance (Fig. 5.2).

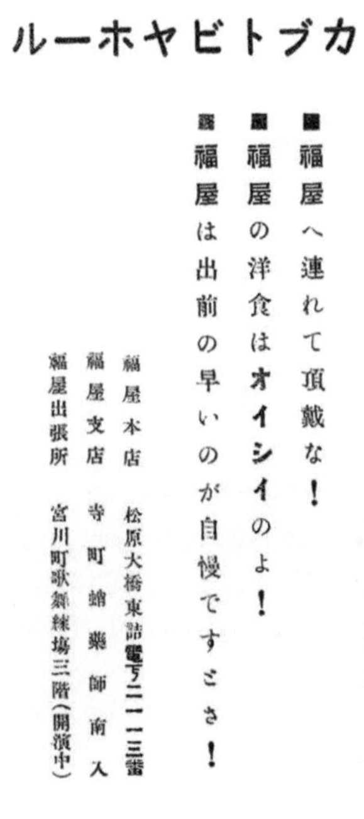

Figure 5.1 Advertisement for Fukuya

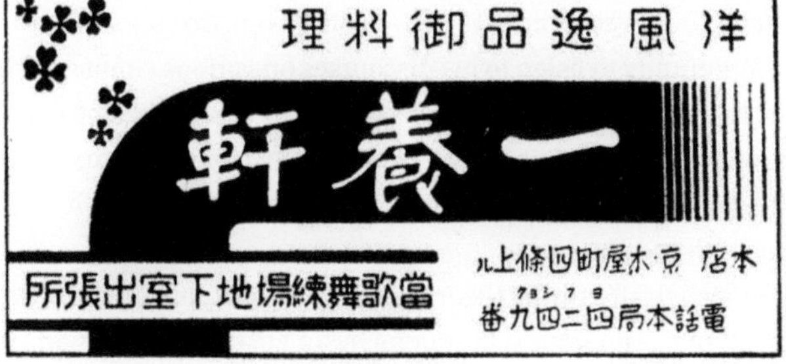

Figure 5.2 Advertisement for Ichiyōken

Fukuya didn't simply open a temporary sub-branch. As indicated by Fukuya's slogan "Fukuya prides itself on its fast delivery!" it also offered a delivery service like Fujiya. This is another characteristic of *iromachi yōshoku* – the Western-style food of the *kagai*.

5.3.3 *Iromachi yōshoku*: Western-style food of the *kagai*

In this section, we will look at a gourmet article that sheds light on *iromachi yōshoku*, the Western-style food that was associated with the *kagai* at that time. As we have seen, Roppa commented on the number of "-yōken" in Kyoto. In 1935, Mikumi Yamakawa published an essay on Goyōken, another "-yōken" restaurant, which was situated on Hanamikōji on the south side of the Gion district.[41] "Mikumi" may be a pen name; he contributed several interesting articles to the first issue of *Rakumi* magazine, but not much is known about him. It seems that he was an acquaintance of Tenmin Matsuzaki (1878–1934), known for a collection of photos of the Ginza district in Tokyo, so it seems that he was in the same industry (i.e., a writer). *Studies of Kyoto Bars* touched on this article, but we will quote it in its entirety here, given its great value as a historical source:

> Goyōken is a popular restaurant offering French cuisine in the heart of the Gion *kagai* district. The food there is a little light on the seasoning. Why is this? The taste of the food remains the same regardless of when one eats it, so it is not the case that one adds seasonings depending on how things have turned out on a given day[.] French cuisine is probably the lightest of Western cuisines, but in addition to the lightness of the seasoning, it leaves a pleasantly sweet sensation on the tongue. I believe that there is some reason for such fineness of seasoning.
>
> It was a clear early afternoon in the rainy season, Hanamikōji street was quiet; our midnight was daytime for the pleasure districts. I called in and presented my business card to the master of the establishment [i.e., Goyōken]. He welcomed me warmly into the saloon. I asked him about the taste that I had experienced at Goyōken. His brow furrowed with an air of intent, he answered with decisive confidence: 'Yes, your taste buds are sensitive. Our business is with the pleasure districts of Gion; we know the tradition-steeped tastes of those districts, and we only create flavors that are suited to those tastes.'
>
> I took this opportunity to listen to his discourses on various culinary philosophies.
>
> The master had been a frequent customer at this establishment ever since it first opened. The former owner had to quit due to unforeseen circumstances, and the erstwhile customer ultimately became the owner. It had happened by chance. And now the place was thriving more than ever.
>
> He had developed the following theory based on that experience. Namely, that taking over management of a restaurant is best done by an amateur rather than an expert, and among the amateurs, someone who has been a customer of the restaurant is the most suitable. This is because, as a customer, they will have developed a critical eye for many

aspects [of the business] over a long period of time. So, the current business style was focused on customers (although it also took four people to work on deliveries). Given the characteristics of the place, there were always multiple customers. Therefore, the young waiters [literally "boys"] had to offer complex and multi-faceted customer service, which seemed very demanding. There were at least six different possible combinations: customers and geishas; customers and maikoes; groups of geishas; geishas and maikoes; maikoes and *nakai* [waitresses or hostesses at traditional Japanese establishments], customers and *nakai*, and so on. The atmosphere of the table differed depending on which combination it was, so the staff would have to offer two, three, or more different kinds of service. To work here, the young waiters must be geniuses with a sixth sense. The master puts the waiters' training first. Not only during business hours but also during the daytime when the girls casually wander in wearing yukata, he tells the waiters that they just can't be in a yukata mood [i.e., relaxed and casual].

The rooms are divided roughly into booths [literally, "boxes"]. A certain amount of erotic-grotesque nonsense can be overlooked as customary in the pleasure district. However, bad-taste tomfoolery of the modern sort is right out. In that respect, the girls take pride in being the most famous geishas in all the land.

My pen is itching to continue its pilgrimage, and so I would like to close here. Try asking any maiko, "is there somewhere around here that serves up a delicious rice curry?" She will surely be tempted to reply, "I like Goyōken's rice curry the best." Such is the reputation that Goyōken's rice curry, along with its sandwiches, has developed among the talkative folk of the pleasure quarters.

Mikumi Yamakawa visited the popular French restaurant Goyōken because he felt drawn in by the light flavor of their food, which was the same no matter when you went. Passing through the deserted Hanamikōji street during the daytime, he was ushered into the saloon of Goyōken, where he could directly question the proprietor.

The proprietor (master) assured him that as long as he was doing business with the customers of the pleasure districts (i.e., *kagai* or *iromachi*) called Gion, he would "only create flavors that are suited" to the "tradition-steeped tastes of those districts." The proprietor tells him that as he himself used to frequent the restaurant as a customer, he was able to nurture it into a thriving business because he had developed an objective, "critical eye for many aspects."

Of course, it is not only the taste of the food but also the nature of the business and service that have been tailored to suit the "characteristics of the place." Like Fujiya and Fukuya, it offered a delivery service; the fact that it required "four people to work on" gives us a glimpse of its popularity. Unlike Roppa's experience described above, the table seats were not separated by curtains, and the booth seating may have been modeled on the style of seating found in Gion Café invented by Ginjirō Takahashi. Although we are told that "bad-taste tomfoolery of the modern sort is right out," such seating may have indeed allowed for this sort of mischief.

In another recollection, we are told the following of Gion Café:

> Sometimes, for example on the way back from the cinema, I would stop by and have a beer. Geishas and maikoes of Gion would be brought in by guests, and I would sometimes see them delicately eating things like fried shrimp with their dainty mouths. *The waitresses wore white aprons and served beer with simple good-natured grace* but did not come and wait on the guests in their seats. In short, it was as if a "Western-style restaurant" had drawn one step closer to being a "café," but I nonetheless enjoyed it a great deal.[42]

Places such as these where the "geishas and maikoes of Gion would be brought in by guests" were an extended space of the *iromachi yōshoku* Western-style cuisine of the pleasure districts, yet a decisive difference from Goyōken was their use of waitresses instead of waiters (which are referred to using the English word "boy" in the original Japanese sources). Being a "boy" at such an establishment would have entailed what Roppa referred to as "the kind of service where they would bend over backward for you." It was a role in which one was required to instantly distinguish between different combinations of customers of virtually unlimited complexity and attend to their needs at all times. The role naturally required training, and the waiters were required to be "geniuses with a sixth sense."

In this sense, *iromachi yōshoku*, including its waitress-service cafés, no longer exists.

5.4 The influx of Western-style restaurants into the *kagai* entertainment districts

5.4.1 Long-established restaurants on the move

As we have seen, a Japanese-style western food culture flourished in and around the *kagai* entertainment districts. Shōtarō Ikenami's beloved Takarafune was one of the restaurants that pioneered this trend. Although it was located on the south side of Shijō-dōri before the war, it had moved to the heart of the geisha teahouse district by the time Ikenami visited. In fact, many Western-style restaurants moved repeatedly, and their choices of location reveal something interesting about their intentions.

For instance, Tōyōtei, the oldest Western-style restaurant in Kyoto, left its original location on the west side of northern Kawaramachi Sanjō, in the late 1930s (early in the Shōwa era), and moved to a location (Minamidaikoku-chō) on the corner in the northern part of the Nishikiyamachi Sanjō intersection and became a tonkatsu (pork cutlet) specialty restaurant (Fig. 5.3). Moving from Kawaramachi-dōri to Nishikiyamachi-dōri meant being closer to the Pontochō *kagai*. In fact, the new location was just around the corner from the Kaburenjō theater, as indicated by the advertisement: "Tonkatsu Tōyōtei, located on your way home from the Kamogawa Odori dance performance."

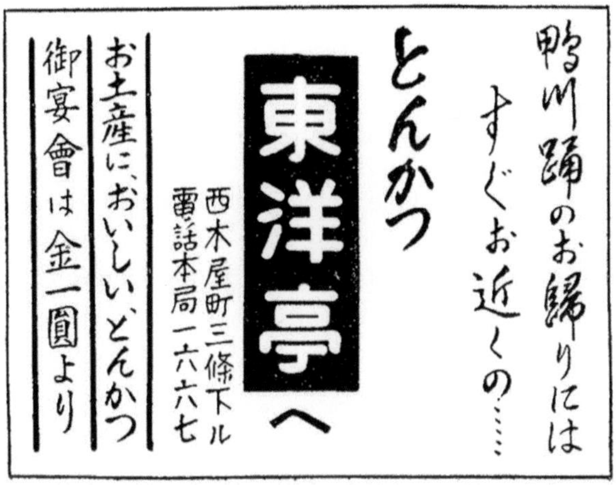

Figure 5.3 An advertisement for Tōyōtei featuring tonkatsu

Tōyōtei moved to its present location in the suburbs of Kitayama in 1966, as if to avoid the hubbub of Nishikiyamachi, which had transformed into a "nightlife district" in the late 1950s. This marked a shift away from its focus on the *kagai*.

Another veteran restaurant, Manyōken, also made a series of interesting relocations. As previously discussed, it initially moved to Shijō-dōri, which was expanded in the early twentieth century, and where it continued in business for nearly 90 years. Then, in 2001, it moved to Shinbashi-dōri at the northern end of the Gion Shinchi *kagai*, before finally moving to its current location on the western corner across from the Kaburenjō theater on the south side of the Gion district. Thus, the venerable French restaurant, which had been the least strongly associated with Roppa's *iromachi yōshoku*, eventually chose to locate itself smack-bang in the middle of Kyoto's most iconic *kagai*.

Kaiyōtei was initially located in the *ishidanshita* area below the stone steps of Yasaka Shrine and has moved back and forth over the Kamo River twice. In 1949, after the war, the restaurant moved to the side of the Kamo in the southern part of Pontochō, and it appears that it will now be relocated again to a new building in the vicinity of Miyagawa-chō back on the other side of the Kamo. Other branches, such as Shin-kaiyōtei ("New Kayōtei") were opened on the south side of the Gion district in 1959,[43] showing a consistent orientation toward the *kagai* (Fig. 5.4).

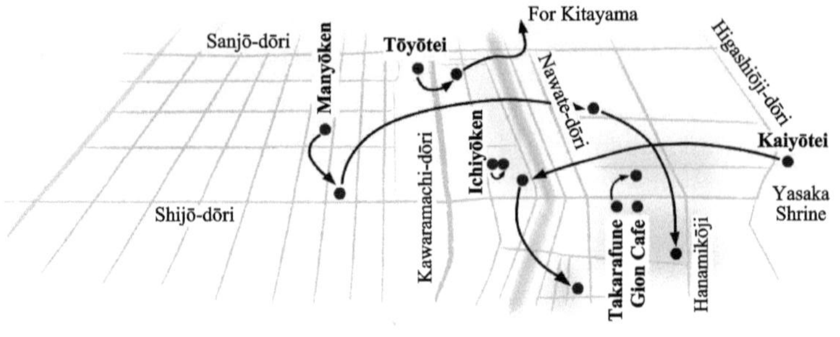

Figure 5.4 The influx of Western-style restaurants into the *kagai*

After the war, in addition to the changes in location of long-standing Western-style restaurants as described above, some restaurants were established by converting former geisha teahouses. Manshun in the Kamishichiken *kagai* was a French restaurant that inherited its name from a teahouse, and with hindsight, can be considered as a pioneer of the "*machiya* French" style (it closed in 2014). The katsu sandwich there was a signature item on the menu.

5.4.2 Beef steak with *ochazuke*

One restaurant that cannot be forgotten when considering the history of Western-style cuisine in postwar Kyoto is Tsubosaka, founded in 1947, which Jinnosuke Inoue called "the best Western-style restaurant in Kyoto." Shōtarō Ikenami wrote, "I felt like eating beef steak, so I went to [Tsubosaka] on Hanamikōji for the first time in a long time," indicating that he had more than a passing familiarity with the place.[44]

Inoue explains, "Because of the *character of the location*, in the middle of the pleasure quarters on the north side of the Gion district, it had no choice but to become adept at catering to discerning customers."[45] Kinosuke Usui pointed out that "It became popular among maikoes and geishas, who would say 'If you eat western food, then Tsubosaka is the place.'" Usui watched the food being prepared while sitting at the counter, and named it "*itamae yōshoku*," combining a word for traditional Japanese high-end food preparation (*itamae*) with the word for Western-style cuisine (*yōshoku*).[46]

After Inoue died, Ayako Kokubu took over the "Tsubosaka (Western-style food)" entry in *Kyoto Taste Map* and she, too, praised the restaurant:

> The croquettes are small, crispy, and have a light texture typical of Gion.
>
> In general, the food is clean, rich, and not greasy, which is also *typical of Gion*, so that even the proprietress of a tea house, geisha, or maiko will come to eat the food there without any hesitation.[47]

This sense of Tsubosaka being "typical of Gion," emphasized in italics above, is something that all three writers mention.

> Finally, the fact that if you so desire, they will provide you with chopsticks and a tea bowl, serve you pickles, or even go to the trouble of making *ochazuke* [tea rice soup], is truly an expression of *Kyoto style and the local character*. (Inoue)[48]
>
> If you want *ochazuke* later, they will serve you pickles, rice, and tea. This way of doing things is also *typical of the Gion district*. (Usui)[49]
>
> At the end, you will receive a light *ochazuke* in a tea bowl, then delicious homemade salted kombu seaweed, pickles, and other items, all served with great attention to detail, including chopsticks and green tea, all of which makes you ever more pleased with its *typical Gion character*. (Kokubu)[50]

Western-style food with traditional Japanese *ochazuke* feels very typical of Kyoto, especially Gion. Ikenami used the words "beef steak," but Grill Tsubosaka's specialty was the "teriyaki set meal" comprising potage soup, *teriyaki steak*, salad, *ochazuke*, frozen dessert, and demi-coffee. The steak was "seared teriyaki with a unique sweet and spicy Japanese-style sauce," with further depth added to the flavor by sprinkling the steak according to the customer's preference with finely chopped onions that had been soaked in water. Today, the popular counter service Western-style restaurant Mishina in Ninenzaka is also known for serving *ochazuke*, but the *iromachi yōshoku* restaurant Tsubosaka was the first establishment to do so.[51]

In 1975, the then owner of Tsubosaka said, "We're not really what you might call a restaurant. It's a family-run place, and we use very basic methods to cook old-fashioned dishes that you could find anywhere, so you could say that we're bit of a fossilized Western-style food joint."[52] This view of cuisine, shared with that expressed by the owner of Takarafune, must have been what drew Ikenami in.

5.4.3 The last *iromachi yōshoku*: Western-style food of the pleasure districts

> But, you know, that feeling that you'd get around places like Ichiyōken in Kiyamachi, listening to the murmuring of the river, eating a bowl of shrimp coquille or something like that with a few drinks, waiting for the geisha to come over, you know, it was just a ridiculously nice classy feeling, pretty different from waiting for your mistress to show up while you poke at some hotpot.[53]

When Roppa Furukawa visited, Ichiyōken faced onto the street of Kiyamachi-dōri, and Takase is not a particularly large river, so you wouldn't have been able to hear the river murmuring inside the restaurant, but he gives us a sense of the atmosphere of Ichiyōken at that time.

In 1978, Ichiyōken moved down an alley to a location that that was all but invisible from a distance (albeit bringing it that bit closer to the Pontochō entertainment district) and has remained in business as a hole-in-the-wall-type Western-style bar-cum-restaurant. It opened in 1922, so it has now been in business for over 100 years.

Seated at the counter in front of the shelves lined with bottles of Western liquor and looking at the wizened à la carte menu, we feel as if we can almost smell that "scent of Western-style cuisine from the early Shōwa period." We can even see that "shrimp coquille" that Roppa perhaps ate (Fig. 5.5).

Figure 5.5 The à la carte menu at Ichiyōken

Seeing items such as "haishi rice" (emphatically not "hayashi rice") on the menu harkens back to Bussankan's menu from 1926. Comparing both Bussankan and Ichiyōken's à la carte dishes (Table 5.5), besides the fact that the price has risen between roughly 20 and 5,000 times, Ichiyōken's contemporary menu appears to be

Table 5.5 Dishes served by both Bussankan (1926) and Ichiyōken (2015)

Bussankan (1926)	Ichiyōken (2015)
Set meal (1 yen)	Set meal (3,030~5,400 yen)
Fried shrimp (50 sen)	Fried shrimp (1,080 yen)
Chicken cutlet (50 sen)	Chicken cutlet (1,300 yen)
Beef cutlet (35 sen)	Beef cutlet (1,730 yen)
Beef steak (50 sen)	Beef steak (3,240 yen)
Chicken sauté (50 sen)	Chicken sauté (1,300 yen)
Mince balls (35 sen)	Mince balls (710 yen)
Ham and eggs (40 sen)	Ham and eggs (920 yen)
Chicken rice (with coffee) (40 sen)	Chicken rice (980 yen)
Haishi rice (with coffee) (35 sen)	Haishi rice (1,520 yen)
Sandwich (40 sen)	Sandwich (870~1,620 yen)
Vegetable salad (25 sen)	Vegetable salad (710 yen)
Ham salad (40 sen)	Ham salad (870 yen)
Asparagus (35 sen)	Asparagus (640 yen)

much the same as Bussankan's was in the Shōwa period. On this basis, it might be safe to assume that Ichiyoken's menu has remained unchanged since the Shōwa period.

There are no "boy" waiters or curtains to partition the tables in this restaurant, which is all but invisible from afar, and yet it remains steeped in the scent of the Western-style cuisine of the pleasure districts from the early days of the Shōwa era – the last of the *iromachi yōshoku*.

Notes

1. Ikenami, Shōtarō, "Shokutaku no jyōkei (rensai dai 29 kai): Gion matsuri sono 1" (Scenes at the dining table (Serial no.29): Gion festival part 1) (*Shūkan Asahi* vol. 77, No. 29, 1972), pp. 66–67. Ikenami, Shōtarō, *Shokutaku no jyōkei* (Scenes at the dining table), Shinchō Bunko, 1980, p. 168.
2. Kokubu, Ayako, *Gensen Kyoto Ajishirube* (Selected Kyoto Ajishirube), Kasandō, 1991, pp. 146–147.
3. *Gekkan Kyoto: 6 gatsu gō* (Monthly Kyoto: June issue), vol. 97, 1987, p. 21.
4. Ikenami,"Shokutaku no jyōkei, sono 1," p. 67.
5. *Gekkan Kyoto: 6 gatsu gō*, p. 21.
6. From Bussankan's "Shinkan goannai" (Guide to the new building) (pamphlet) owned by the authors.
7. *Kyoto Nichinichi Shimbun*, September 5, 1928.
8. Kawaru, Tsūko "Kyoto no inshoku ten" (Restaurants and bars of Kyoto) (Sasa Masaichi ed., *Yoru no Keihan* (Keihan at night) Vol. 16, Kinkōdō, 1903), pp. 63–69. Quote from p. 65.
9. Tanizaki, Junichirō, "Suzaku Nikki (2)" (Suzaku diary, vol. 2) (*Osaka Mainichi Shimbun*, April 28, 1912).
10. For more on this point, see: Kato, Masahiro, *Sakaba no Kyoto gaku* (Studies of Kyoto bars) (Minerva Shobō, 2020), Chapter 3.
11. Teikoku Kōshinsho, Kyoto Shisha, ed., *Kyoto shō kō taikan* (Encyclopedia of Kyoto merchants and artisans), Teikoku Kōshinsho, Kyoto Shisha, 1928, p. 269. The website Capital Tōyōtei states that, on May 1, 1897 "The Tōyōtei Hotel will open in Kyoto and offer Western food for the first time," so it seems that it was several years earlier than this passage. Capital Tōyōtei website (http://www.touyoutei.co.jp/company/history/), last viewed on January 5, 2021.
12. Kyoto Central Telephone Bureau, ed. *Kyoto denwa bangō bo* (Kyoto telephone directory) (Kyoto Central Telephone Bureau, 1938, p. 72) has an entry for "Kaiyōtei Shiten, Bessho Tsuneichi, southern Karasuma, Shijō, Western cuisine."
13. Nippon Telegraph and Telephone Public Corporation Kinki Telecommunications Bureau ed. *Kyoto-shi shokugyō betsu denwa bangō bo* (Kyoto city telephone directory by occupation) Nippon Telegraph and Telephone Public Corporation Kinki Telecommunications Bureau, 1964, p. 753.
14. *Kyoto Nichinichi Shimbun*, July 24, 1926.
15. *Kyoto Nichinichi Shimbun*, July 17, 1929.
16. Tanizaki, "Suzaku Nikki (2)."

17. Planning Thirteen, ed., *Ranteibō*, Manyoken 75th Anniversary Publishing Editorial Office, 1985.
18. *Kyoto Nichinichi Shimbun*, June 30, 1926 (evening edition).
19. For further details, see: Kato Masahiro and Kawasumi Naomi, "Kindai Kyoto ni okeru shuyō shōtengai no tenpo fukugen: 'Gion-machi' o jirei toshita hōhō no kentō" (Restoration of the main shopping streets in modern Kyoto: Examining methods using the case of 'Gion-machi') (*Rekishi chirigaku* (Historical geography), vol. 62 no. 4, 2020), pp. 1–17.
20. Kishimoto, Suifu, "Kei Han Shin sakariba fūkei" (Scenes of the amusement districts of Kyoto, Osaka, and Kobe) (Sakai, Masato and Kishimoto, Suifu, *Santo sakariba fūkei* (Scenes of the amusement districts in the three cities), Seibundō, 1932), p. 87.
21. Tsuji, Fuku, "Shijō seki to Nawate" (Shijō seki and Nawate) (Osaka Mainichi Shimbun, Kyoto branch ed., *Kyoto shin 100 kei* (100 New Views of Kyoto), Shinjidaisha, 1930), pp. 105–109. Quote is from p. 108.
22. Saitō, Hikaru, *Maboroshi no 'kafē' jidai: Yoru no Kyoto no modanizumu* (The Age of the Phantom Café: Modernism in Kyoto at Night), Tankōsha, 2020, pp. 64–69.
23. Yamamoto, Shūji, "Nomi aru ki: Dai 13 shō" (Drinking and walking diary: Chapter 13) (*Rakumi*, vol. 25, 1940), pp. 73–75. Quote from p. 73.
24. Maeshiba, Kenzō, "Shun shō yomoyama banashi: Issedai mae no Kyoto o shinobi tsutsu" (Spring Festival Yomoyama Stories: Reminiscent of Kyoto a Generation Ago) (*Rakumi* vol. 66, 1956), pp. 21–24. Quote from p. 23.
25. Kawai, Yoshishige, "Kyoto ryō in 10 nen shi" (10-Year History of Kyoto Food and Drink) (Kyoto Ryōin Shimbun, 1970, p. 209), secondary quoting an article in *Kyoto Ryori Shimbun* (February 1, 1936).
26. *Rakumi*, No. 2, 1935, p. 53.
27. Ōkubo, Tsuneji, *Umaimon jyunrei* (Pilgrimage of delicious things), Rokugatsusha, 1958, pp. 40–41.
28. Ōkubo, Tsuneji et al., eds., *Kyoto, Osaka, Kobe: Umai mise 200 ten* (Kyoto, Osaka, Kobe: 200 delicious restaurants), Shibata Shoten, 1964, p. 225.
29. Furukawa, Roppa, "Roppa Shokudan [4]" (*Amakara* No. 21, 1953), pp. 20–22. Quote from p. 22.
30. Furukawa, Roppa, "Roppa Shokudan [44]" (*Amakara*, No. 65, 1957), pp. 51–53. Quote from p. 51.
31. This series is summarized in *Roppa Shokudan: Kanzen ban* (Roppa Shokudan complete edition) (Kawade Bunko, 2014).
32. Furukawa, Roppa, "Roppa Shokudan [7]" (*Amakara*, No. 25, 1953), pp. 11–13. Quote from p. 11.
33. Furukawa, "Roppa Shokudan [44]," p. 51.
34. Furukawa, p. 52.
35. Furukawa, p. 52.
36. Kokubu, Ayako, *Zoku, Kamera: Kyō aji 100 sen* (100 selections of Kyoto flavors: Camera edition), Tankō Shinsha, 1964, p. 68. For the identification of the location, please refer to Kato Masahiro, *Sakaba no Kyoto gaku* (Minerva Shobō, 2020, p. 94).

37. Yoda, Yoshikata, "Kamigata mikaku chizu 1: Ōtō" (Kamigata taste map 1: Ōtō) (*Tankō*, vol. 14, no. 1, 1960), pp. 66–69. Quote from p. 69.
38. *Shūkan tomato & terebi Kyoto* (Weekly tomato & tv Kyoto), January 3, 2014.
39. *Asahi Shimbun*, March 16, 2017.
40. Satō, Junkichi and Hirano, Kumazō, eds. *Kyoto jitsugyō kai* (Kyoto business world), Hakushinsha, 1912.
41. Yamakawa, Mikumi, "Aji to kibun o tazunete (3)" (Visiting taste and mood (3)) (*Rakumi*, vol. 1, no. 3, 1935), pp. 87–93. Quote from pp. 89–90.
42. Maeshiba, "Shun shō yomoyama banashi," p. 23.
43. Research Department, Kyoto Chamber of Commerce and Industry, ed., *Kyoto shō kō jinmei roku; Shōwa 39 nen ban* (Records of famous Kyoto merchants: 1964 edition), Kyoto Chamber of Commerce and Industry, 1964, p. 449.
44. Ikenami, *Shokutaku no jyōkei*, p. 87.
45. Inoue, Jinnosuke, "Tsubosaka: Yōshoku" (Tsubosaka: Western food) (Sōgensha Editorial Department, ed., *Kyoto mikaku chizu: 1969 nen ban* (Kyoto Taste Map 1969 Edition) Sōgensha, 1968), pp. 50–51.
46. Usui, Kinosuke, *Shinpen, Kyoto mikaku sanpo* (A gustatory stroll through Kyoto: New edition), Shirakawa Shoin, 1970, p. 123.
47. Sōgensha Editorial Department, ed., *Kyoto mikaku chizu: 1975 nen ban* (Kyoto taste map 1975 edition), Sōgensha, 1975, p. 77.
48. Sōgensha Editorial Department, *Kyoto mikaku chizu:1969*, p. 51.
49. Usui, *Shinpen, Kyoto mikaku sanpo*, p. 124.
50. Sōgensha Editorial Department, ed., *Kyoto mikaku chizu:1975*, p. 77.
51. *Gekkan Kyoto: 6 gatsu gō*, pp. 16–17. Kiyoji Tsubosaka, the owner at the time proudly called Toshiaki Mishina "my son" and named him his heir.
52. *Gekkan Kyoto: 6 gatsu gō*, p. 16.
53. Furukawa, "Roppa Shokudan [44]," p. 52.

10.2478/9784991351600-013

Chapter 6
From Culinary Sacred Grounds

6.1 The *Tamagaki* sacred fence of Yamakage Shrine

At the southeast corner of the intersection between Higashiyama and Higashi-ichijō roads stands a pillar indicating the approach to the Yoshida Shrine, where one can find the Yamakage Shrine. Although the eastern end of Higashi-ichijō (to the east of Higashiōji-dōri) is an old road, there is nothing to suggest the approach to a shrine, perhaps because the road is sandwiched between two campuses of Kyoto University (the former Kyoto Imperial University) to the north and the former Daisan Kōtō Gakkō ("Third High School") to the south. At the end of this road is a large vermilion-lacquered torii gate, after which the road turns into the gravel approach to the shrine that doubles as a place for devotees to park their bicycles.

After the second torii gate, there is a stone stairway lined with rows of stone pillars that look like a *tamagaki*. *Tamagaki* are stone fences surrounding a Shinto shrine, inscribed with the names of people and organizations who donated money, often to shrines of their associated industries. It goes right up to the highest point, on the strip of land that divides the steps from the roadway to the right, with a sign stating that they are a "Dedication of the Daisan Kōtō Gakkō Alumni Association" (November 1977). The alumni include a famous geographer (Saneshige Komaki).

The main building of Yoshida Shrine itself is located on the left (the north side of the precinct) after climbing the stone steps. The climb then continues up a fairly steep slope that extends south from the torii gate of the main shrine, soon coming to a fork that branches off to the approach to the Kaso Shrine, which enshrines the god of sweets and confectionary. We continue along our way. On a slightly elevated spot of flat land that seems to have been leveled out of the western slope of Mount Yoshida, we find Yamakage Shrine. This shrine preserves the spirit of Fujiwara no Yamakage, a Heian period court noble and founder of the Shijō school of *hōchōshiki* (the "knife ceremony" – a ritualized culinary practice using a special knife and chopsticks to fillet fish, chicken, etc. without directly touching it), who is connected with Yoshida Shrine and is venerated as the ancestor of the culinary industry (Fig. 6.1). The shrine was established on May 8, 1959, by bringing together people involved in the food and beverage industry in Kyoto, spearheaded by Fujinosuke Kitamura of Hiranoya.[1]

Yamakage Shrine's *tamagaki* stone fence comprises 77 stone pillars that surround the shrine precinct. The *tamagaki* was constructed in May 1969 to commemorate the 10th anniversary of the shrine. The stone pillars are densely engraved with the names of organizations and restaurants, including those of the local Kyoto food and beverage industry (Fig. 6.2 and Table 6.1). Only a few organizations' names are inscribed there, yet they include organizations from across Japan, from Hokkaidō and Tōhoku to Kagawa Prefecture. It is worth noting the districts where the 14 organizations from

Figure 6.1 Yamakage Shrine at Yoshidayama

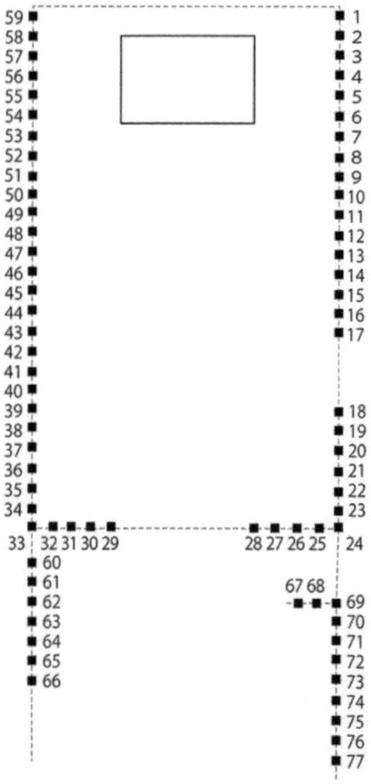

Figure 6.2 The *tamagaki* (arrangement of stone pillars) at Yamakage Shrine

Tokyo are located – Yoshichō, Akasaka, Ginza, Shibuya, Yanagibashi, and Asakusa – many of which are associated with *kagai* entertainment districts (referred to as *sangyōchi* "district of the three businesses" – referring to entertainment districts with teahouses, restaurants, and geisha establishments). We also see donations from the cities of Shimizu in Shizuoka Prefecture and Handa in Aichi Prefecture, both within the Chūbu region. The Kyoto organizations listed include the Kyoto Western Food Association (established in 1955), as well as food and beverage industry associations for the different city regions (Kawabata, Shichijō, Higashiyama, and Gojō).

Table 6.1 Engraved on the *tamagaki* at Yamakage Shrine

1.	Sapporo Japanese Cuisine Chefs Association Federation of Fukushima Japanese Cuisine Associations Federation of Yamagata Food and Beverage Industry Associations
2.	Sapporo Japanese Cuisine Restaurant Cooperative Association
3.	Yamagata Japanese Cuisine Chefs Association
4.	Kagawa Japanese Cuisine Chefs Friendly Society
5.	Tokyo Shibuya Japanese Cuisine Business Association
6.	Tokyo Nagatachō Cuisine Association
7.	Manshū, Inokuma Kamichōjamachi
8.	Kanemasa, southern Nawate Sanjō
9.	Kamaboko Ibarakiya, Yanagibaba Oshikōji
10.	Tsuruse, Kiyamachi Gojō
11.	Kyō-ryōri Minokichi, Awataguchi
12.	Isechō, Shinmachi Nishiki
13.	Manmata, Nishijin
14.	Matsutomo, Pontochō
15.	Tsuruya Shiten, Shimizu
16.	Fujiya, Kibune
17.	Kyoto Kawabata Restaurant Association Kyoto Shichijō Restaurant Association
18.	Kyoto Higashiyama Restaurant Association Kyoto Western Food Association
19.	Uoshin, Nishijin
20.	Tsuruya, Okazaki
21.	Sōhonke Kawamichiya
22.	Kawamichiya Noren Association
23.	Kawamichiya Yōrō, Shōgōin
24.	Shimizu City Cuisine and Restaurant Association Asahikawa Japanese Cuisine Chefs Association Aichi Handa Cuisine and Restaurant Industry Association Noshiro City Restaurant Association

(Continued)

Table 6.1 *(Continued)*

25.	**Kandagawa, Myōjinshita**
26.	**Kandagawa, Dōjima**
27.	**Kandagawa, Saiseki**
28.	Kyoto Federation of Food and Beverage Industry Associations
	Kyoto Gojō Food and Beverage Association
29.	Tokyo Japanese Restaurant Association
	Federation of Hokkaidō Restaurant Business Associations
30.	Akira Ogawa, Kyoto Culinary School
31.	Hyōtei, Nanzenji
32.	Hiroya, Kibune
33.	Rinji, Nishiki
	Yaomasa, Shinmachi Bukkōji
	Togetsutei, Arashiyama
	Nijō Club, Kiyamachi
34.	Heihachi, Yamabana
35.	**Shin Miura, Kiyamachi**
36.	**Shin Miura, Fukuoka**
37.	**Shin Miura, Gion**
38.	Toriiwa, Higashinotōin
39.	Ryozen Shin Onsen, Cuisine Ryokan
40.	Kinshin, Shinmachi Rokkaku
41.	Soft-shell Turtle Cuisine Daiichi, Nishijin
42.	Kitō, Nishijin
43.	Yoshida Villa
44.	Funatsuru, Kiyamachi Matsubara
45.	Higashiyama Kodaiji Temple Doi Sadae Doi
46.	Iyomata, Nishiki
47.	Ran Iwao Yanai, Shimabara
48.	Minokō, Gion Shimogawara
49.	Kikusui, Nanzenji
50.	Kyō-no-ajimisekai
51.	Nishida Teiichi Chikaramochi, western Shichijō Karasuma
52.	Tokyo Asakusa Three Businesses Association
53.	Tokyo Yoshichō Association
54.	Tokyo Ueno Japanese Restaurant Association
55.	Tokyo Unagi Kabayaki Merchant Association
56.	Tokyo Yotsuya Japanese Restaurant Association
57.	Tokyo Kyobashi Japanese Restaurant Association
58.	Tokyo Kanda Restaurant Association
60.	Tokyo Shimbashi Association

Table 6.1 *(Continued)*

61.	Tokyo Akasaka Restaurant Association
62.	Tokyo Yanagibashi Association
63.	Tokyo Ginza Japanese Cuisine Association
64.	Asahikawa Cuisine Association
65.	Asahikawa Kagetsu Tourism Co., Ltd. President Akiyasu Watabe
66.	Federation of Osaka Culinary Associations
67.	Taiwa Culinary College
68.	Beefsteak Suehiro
69.	Cantonese Cuisine Shin Hamamura
70.	Ginsui, Saiseki
71.	Kikunoi, Maruyama Makuzugahara
72.	Tankuma Minamiten, Kiyamachi Bukkōji
73.	Okimasa, Nishinokyō, Enmachi
74.	Tenki, Kyoto Nishijin
75.	Ōedo, Shijo Takakura
76.	Ōedo, Nishijin
77.	Umemura, Kiyamachi

Notably, however, only restaurants within the city of Kyoto made individual donations. These include many of the restaurants discussed in this book: Kanemasa, Minokichi, Tsuruya, Uoshin, Kandagawa, Hyōtei, Heihachi, Shin Miura, and Daiichi. The reader will likely understand the significance of the three stone pillars of Shin Miura (Gion, Fukuoka, and Kiyamachi) standing alongside each other at numbers 35 to 37 (Fig. 6.3).

Figure 6.3 "Shin Miura" engraved on the stone pillars of the *tamagaki*

Then, what of the Kandagawa pillars at numbers 25 to 27 (Fig. 6.4)?

Figure 6.4 "Kandagawa" engraved on the stone pillars of the *tamagaki*

6.2 Secret sauce

When Jinnosuke Inoue visited Kandagawa to find information,[2] he might have seen the following slogan advertising Kandagawa:

> 3 years for peaches and chestnuts, 8 years for persimmons, 70 years for Edo-yaki in Kyoto

In Chapter 1, we saw that Kandagawa in Myōjinshita opened a branch in Saiseki. However, the *tamagaki* at Yamakage Shrine names another establishment "Kandagawa, Dojima" between the names of the main restaurant and the branch. What is the significance of this? According to Toshirō Kandagawa:

> My father ran an eel restaurant [Kandagawa] in Kyoto. Having a restaurant named Kandagawa in Kyoto may seem a bit strange, but there is a reason for that.
>
> My family originally came from Gunma. My grandfather was the mayor of Saru County in Gunma Prefecture and was later nominated as Governor of Gunma Prefecture. Grandfather refused this position and left for Tokyo.
>
> Grandfather went to Tokyo, bought a plot of land of 3,000 *tsubo* [about a hectare] in Hongō. He asked an eel restaurant [Kandagawa] in Kanda Myōjinshita about the management of the land.

> I don't know why he asked for an eel restaurant, but with its cooperation, the result was that grandfather's eel restaurant [Kandagawa] was born in Hongō.[3]

The reader may notice the error in this explanation and in the continuation of the story. Toshirō Kandagawa (real name Toshirō Ōtake, 1939–2021) often appeared in the competitive cooking program Iron Chef, which was popular in Japan in the 1990s and 2000s. After training at a restaurant in Osaka, Toshirō opened a restaurant in the Ohatsutenjin district in 1965, took over the name of the restaurant, Kandagawa, under the banner of "New Japanese Cuisine" and "Soft-shelled Turtle" instead of eel.[4] This is how the Kandagawa in Dōjima was established.

One thing to add is that the name of the Myojinshita Kandagawa is not derived from the name of the Kanda River. The founder, Moshichi Kanda, was the second son (childhood name: Yasujirō) of Yasubei Udagawa, a lumber merchant in Mukōjima. He was later adopted by a doctor whose surname was Kanda, and after the death of his adoptive father, he reportedly adopted the restaurant name Kandagawa, a combination of the surnames of both families (Udagawa and Kanda).[5]

It is said that every eel restaurant has a pot filled with a secret sauce that has been handed down for generations, and there is story relating to how this sauce was handed down from the first generation at Myōjinshita Kandagawa.

> The amount of sauce is the most important thing for eel. The Kandagawa sauce was made by the owner five generations ago when he first opened the restaurant, and it has been kept the same to this day. The sauce decreases as one uses it, but the old sauce is added to every day, so it is now syrupy and almost oily in the bottle. That is why the sauce is Kandagawa's greatest treasure; if the owner should ever hear that there is a fire, he would go in to get the bottle of the sauce before anything else.[6]

Did Toshirō Kandagawa's grandfather Shigekata Miyata receive an allocation of this sauce from Moshichi Kanda when he opened the Kandagawa branch in Shinsentei? Was the original Kyoto sauce created in the downtown *shitamachi* areas of Edo and then brought to Kyoto 80 years ago?

6.3 Kikusui on Hanamikōji: What happened next

Aside from Yamakage Shrine, even in small shrines that do not appear on maps, you can sometimes see *tamagaki* with the names of historical restaurants carved into them. An example is the Ikuyo Inari Daimyōjin Shrine, which is located along Ogawa-dōri, heading north from Sanjō-dōri. The *tamagaki* faces Ogawa-dōri and features the names of some of the restaurants mentioned in this book, such as Izuu in Gion, Tōyōtei on Kawaramachi, Chimoto in Saiseki, Yaomasa on Shijō, and Hasshin on Yanagibaba.

It would be interesting to find out more about the history of this shrine, but let us finish this book by introducing an additional *tamagaki* in one last place. From the foot of Danguri Bridge, head east on Danguri-dōri toward Gion and pass Yamatoōji-dōri, and you will see a large pillar with an inscription reading "Dakini Sonten" (the name of a Buddhist deity). This is Kōun-an, a sub-temple located at the northwestern end of the Kennin-ji Temple precinct. Kōun-an has a *tamagaki* facing the street, which is rare for a Buddhist temple, perhaps because it enshrines Dakini Sonten, who is also associated with the Shintō deity Toyokawa Inari.

Several of the names of the donors carved into the stone are intriguing, but I let out a cry of surprise when I spotted the words "Kikusui, Hanamikōji" there (Fig. 6.5). This is the restaurant where Junichirō Tanizaki, ate "tori-nabe" (chicken *sukiyaki*) when he first visited Kyoto in 1912. Kikusui no longer exists; its name engraved here is all that remains.

Figure 6.5 The Dakini Sonten *tamagaki*

Among the many sources used in writing this book, none have offered as rich and valuable a description as Tanizaki, except perhaps Mikumi Yamakawa's article reporting on Goyōken (Chapter 5), according to which, in 1935, "Kikusui is a long-established chicken restaurant, in Gion's Hanami district, which had previously created a miso-simmered dish that impressed gourmets, and *which is now inviting chefs from its original home in Nagasaki to start creating pure Nagasaki cuisine.*"[7] Yamakawa lists the menu for Kikusui's set of Nagasaki dishes (2 yen):

> Fins: sea bream fillets, greens, sea cucumbers, half a *kin* [300g] of shrimp, shiitake mushrooms
>
> Large dish: light kudzu, bean dumplings
>
> Miso: sea bream tofu, mustard

Medium dish: Dongpo meat
- Sea bream sashimi, *bōfu* [*Saposhnikovia divaricata*], *bakutai*, wasabi
- *Yubiki* [poached fish], mitsuba, *yoriudo* [vegetable peels]
- *Amago* [trout], *nanbanzuke* [marinated fried salmon]
- Rice, pickles
- *Mizumono* [water-based desert]
- Sweet porridge

All of the side dishes are given above.

Among them, what is referred to as Dongpo meat is cubed braised pork belly. This is a masterpiece of Nagasaki cuisine based on the oral tradition of Su Dongpo... The owner of Kikusui has already attained a degree of expertise in Nagasaki...

It is truly priceless to spend an evening of drizzling spring rain at the red round table in the annex, which is a favorite for tea ceremonies, chatting with friends, drinking sake, and feeling that one is in a different country.[8]

Two decades after Tanizaki's visit, the tatami room was still in the "Irikin style" (an "annex, which is a favorite for tea ceremonies"). Why did Kikusui, specializing in chicken cuisine, adopt Nagasaki cuisine? There are other similar cases, so it seems that chicken and Nagasaki cuisine were a good fit for each other. As we saw in Chapter 4, there is reason to believe that Hakata's chicken *mizutaki* might have originated in Nagasaki (Chapter 4), but in fact, there was a boom in Nagasaki cuisine in Kyoto at the time, and thus Kikusui pivoting to this cuisine was likely a response to the preferences of the times. It may now be too late to know where Kikusui, with its chicken cuisine (Nagasaki cuisine), was located. Tanizaki's writings and the inscription on the Dakini Sonten *tamagaki* indicate that it was located on Hanamikōji. However, it seems that what was then referred to as "Hanamikōji" sometimes included the entire south side of the Gion district, so for many years it was not possible to identify its precise location, but eventually, its location was revealed in an unexpected way.

The origins of *Shippoku* cuisine have been heavily influenced by Chinese cuisine brought over from China. The dishes are called things like small side, general side, large side, and *sume*. Dishes such as dim sum made by removing the acidity from pickled plums and dipping them in batter, kenchin soup, Japanese chicken with carrots, steamed glutinous rice, sesame tofu, tilefish natto, are served in large bowls, and finally *oshiruko* [sweet adzuki bean porridge] is served. Another interesting aspect of this cuisine is that each person gets a little stylish table like a sutra desk, and small plates or bowls are taken from the drawer. *The large premises are scattered with single-story buildings*, and there is a counter near the entrance, where the eldest son who trained at Hyōtei devotes most of his time. His young wife, used to be a maiko, is popular and well regarded as a waitress.[9]

Beginning with a brief explanation of "*shippoku* cuisine," which refers to dishes mainly of Chinese origin that are found in Nagasaki, and looking at the description of "large premises ... scattered with single-story buildings," it becomes clear how Kikusui grew into a full-fledged Nagasaki restaurant from its origins in chicken cuisine. The revised version of the book quoted above also describes the premises as "a large site and gardens, scattered with tatami room buildings," which are still in the "Irikin style."[10]

However, this is not really an article about Kikusui; instead, it provides a description of Toriimoto. As the name Toriimoto (literally, "at the base of the torii") suggests, the restaurant was located in front of the torii of Yasaka Shrine.

> According to the restaurant's history, it was founded in the Kyōhō era [1716–1736], and the current generation is said to be the seventh generation. Its name, Toriimoto, is said to originate from its original location near a torii gate in Shimogawara, Gion. However, *it is now located on the south side of Gion*, on large premises with a garden, with the Gion Kaburenjō theater building visible immediately on the south side.[11]

Thus, Toriimoto was a long-established restaurant that was founded in the Kyōhō era, which had moved to the south side of the Gion district. Indeed, there are various recollections of it, including:

> In Shimogawara, Toriimoto, which is now on Hanamikōji, boasted an old-fashioned building and offered *Gion cuisine*, inspired over a long period by the *shippoku* cuisine of Nagasaki. *This building on Hanamikōji had once housed Kikusui.*[12]

Here we see that it was Toriimoto that had succeeded Kikusui, the pioneer of the south side of the Gion district, although the reason for this is unclear. The August 1938 *Rakumi* features an advertisement that reads, "Gion Cuisine, Kaiseki Cuisine, Toriimoto, Hanamikōji, Gion, Kyoto City" and similarly notes that "It [Toriimoto] has moved to Hanamikōji," so there can be no doubt regarding the restaurant's location.[13]

That being so, we may wonder what exactly is "Gion cuisine".

> Gion cuisine is slightly different from Kyoto cuisine. In the early days, the owner went to Nagasaki, encountered *shippoku* cuisine there, brought back its style, and rearranged it by adding in some Kyoto style.[14]

"Gion cuisine" adds "some Kyoto style" to the *shippoku* cuisine of Nagasaki. This story is reminiscent of how Hiranoya (which, like Toriimoto, was founded in the Kyōhō era) brought back Chinese taro from Nagasaki and developed *ebi-imo*, thereby giving rise to the Kyoto specialty of *imobō* (Chapter 1).[15] However, what draws our attention here is the sequence of events by which Toriimoto (which mainly serves Nagasaki-derived

Gion cuisine) took over the land and buildings of Kikusui, which had switched to focusing on Nagasaki cuisine.

Ayako Kokubu, who has been as influential to the authors as Kinosuke Usui, who first described "Gion cuisine" as a "Kyoto style" addition to "*shippoku* cuisine." Kokubu was deeply familiar with the city of Kyoto and the culture of its cuisine and restaurants. Her legacy includes several commentaries that at times evoke the "power of place," as described in human geography.

For example, referring to the Western-style restaurant Tsubosaka (Chapter 5), she wrote:

> The Gion district is a peculiar place, and while I believe that the establishments that are best suited to its atmosphere are Japanese-style places such as *ryōtei*, *itamae*, and sake bars, quite a few Western and Chinese-style restaurants are doing a good trade. However, curiously enough, many of the restaurants would make you think that Gion had its own style of Western food, or its own style of Chinese food.
>
> In other words, everything becomes 'Gion-ized.' Or to put it a different way, if any restaurant is located in Gion, it will be end up taking on the character of Gion in no time at all.[16]

It's not just what Roppa Furukawa calls "*iromachi yōshoku*" – the Western-style food of the pleasure quarters. *Chinese Food of Kyoto* by Jiang Shangmei is an excellent theoretical discussion of Kyoto's food culture, which clearly demonstrates that not only the "Chinese-style food of Gion" but also the "Chinese-style food of Kyoto" may appropriately be called "*kagai* Chinese food."[17]

Let us here quote from Ayaka Kokubu again:

> Chicken *mizudaki*, which was originally a specialty of Hakata in Kyushu, was brought to Kyoto and became assimilated into the Kyoto style in no time at all. *Kyoto is a place that has the ability to absorb one good thing after another and digest them into the Kyoto style.*[18]

This comment refers to *mizudaki*, which was covered in Chapter 4. When a culture of traveling takes root in a region, things are often adapted without looking back on the journey, or are adapted strategically, so that its roots (origins and routes) are often forgotten, and it becomes indigenous. *Mizudaki* is a good example of this.

This book has traced the genealogy of *imobō*, freshwater and saltwater fish cuisine, eel *kabayaki*, chicken *mizudaki*, and Western-style cuisine. There will undoubtedly be aspects of the diverse modern food landscape that we have overlooked. Even if we limit ourselves to Kyoto, it seems that there are many aspects of food and restaurant culture that still need to be explored.

This journey of exploration will continue.

Notes

1. Kyoto Federation of Food and Beverage Industry Associations, *Kyoto ryō in 70 nen enkaku shi* (Kyoto food and beverage 70-year history), Kyoto Federation of Food and Beverage Industry Associations, 1993, pp. 58–59.
2. Inoue, Jinnosuke, "Kandagawa (unagi)" (Kandagawa (eel)) (Sōgensha Editorial Department, ed. *Kyoto mikaku chizu* (Kyoto Taste Map), Sōgensha, 1964 [Colophon: *1964 nendo ban Kyoto mikaku chizu* (Kyoto taste map 1964 edition)], pp. 122–123. Quote from p. 122. In the photo on the same page, there is a three-section *noren* curtain bearing the words "Tokyo Unagi Kandagawa."
3. Kandagawa, Toshirō, *Manaita hansei ki* (Diary of a life at the cutting board), Taru Shuppan, 1995, p. 69.
4. Kandagawa, *Manaita hansei ki*, p. 108.
5. Nagai, Kikuji, ed., *Nihon shōzō taikan dai 1 kan* (Compendium of Japanese portraits, vol. 1), Yoshikawa Kobunkan, 1908, p. 5.
6. Shimizu, Seifū (Nakamura Kaoru, ed.), *Kanda no densetsu* (Folklore of Kanda), Kanda Kōronsha, 1913, pp. 106–107.
7. Yamakawa, Mikumi, "Aji to kibun o tazunete (2)" (Visiting taste and mood (2)), *Rakumi*, vol. 1, no. 2, 1935, pp. 75–82. Quote from p. 82.
8. Yamakawa, "Aji to kibun o," p. 82.
9. Kokubu, Ayako, *Shintei ban Kyoto Ajishirube* (Kyoto Ajishirube, New revised edition), Shinshindō, 1980, p. 56.
10. Kokubu, Ayako, *Shin Kyoto Ajishirube* (New Kyoto Ajishirube), Shinshindō, 1984, p. 56.
11. Kokubu, *Shintei ban*, p. 56.
12. Dōmoto, Kansei, "Meiji no koro no Kyoto ryōri ya" (Kyoto restaurants in the Meiji era), (*Rakumi* vol. 100, 1960), pp. 34–36. Quote from p. 36.
13. *Rakumi* vol. 4 no. 3, 1938, p. 131; *Rakumi* vol. 4, no. 4, 1938, p. 47.
14. Kokubu, *Shintei ban*, p. 56.
15. Usui, Kinosuke, *Kyoto mikaku sanpo* (A gustatory stroll through Kyoto), Shirakawa Shoin, 1962, p. 24.
16. Sōgensha Editorial Department, ed. *1975 nen ban Kyoto mikaku chizu* (Kyoto taste map 1975 edition), Sōgensha, 1975, p. 76.
17. Jiang, Shangmei, *Kyoto no Chūka* (Chinese food in Kyoto), Genōsha Bunko, 2016.
18. Kokubu, Ayako (text), Hamabe, Kiyoji (photograph), *Kamera Kyō aji 100 sen* (100 selections of Kyoto flavors: Camera edition), Tankō Shinsha, 1963, p. 80.

10.2478/9784991351600-014

Afterword

Kinosuke Usui was quoted in Chapter 2 as saying: "I always say that *imobō*, mackerel sushi, and herring soba are representative of the true taste of Kyoto, which cannot be found anywhere else." Regardless of whether these dishes fall under the category of Kyoto cuisine, they are nonetheless truly representative of Kyoto. The adroit combination of *ebi-imo* from Nagasaki, cod and herring from Hokkaidō, and mackerel from Wakasa have been combined to create the distinctive dishes of Kyoto.

In writing this book, we always had Kinosuke Usui's *A Gustatory Stroll through Kyoto* and Ayako Kokubu's *Kyoto Ajishirube* to hand. Even after we finished writing this book, we continued to make new discoveries as we leafed through the pages of these works. We hope that our readers will join us in taking advantage of any opportunities to take a stroll through the taste map in future adventures.

November 2021
Masahiro Kato and Naomi Kawasumi

10.2478/9784991351600-015

Index

Index of Personal Names

Index of Place Names

Bei Fragen zur Produktsicherheit wenden Sie sich bitte an:
If you have any questions regarding product safety,
please contact:

Walter de Gruyter GmbH
Genthiner Straße 13
10785 Berlin
productsafety@degruyterbrill.com